Also by Jay Ingram

The Science of Why: Answers to Questions About the World Around Us
The End of Memory: A Natural History of Aging and Alzheimer's
Fatal Flaws: How a Misfolded Protein Baffled Scientists and Changed the Way We Look at the Brain
Theatre of the Mind: Raising the Curtain on Consciousness
Daily Planet: The Ultimate Book of Everyday Science
The Daily Planet Book of Cool Ideas: Global Warming and What People Are Doing About It
The Science of Everyday Life
The Velocity of Honey: And More Science of Everyday Life
The Barmaid's Brain and Other Strange Tales from Science
The Burning House: Unlocking the Mysteries of the Brain
A Kid's Guide to the Brain
Talk Talk Talk: Decoding the Mysteries of Speech
It's All in Your Brain
Real Live Science: Top Scientists Present Amazing Activities Any Kid Can Do
Amazing Investigations: Twins

The Science of Why²

Answers to Questions About the Universe, the Unknown and Ourselves

Jay Ingram

PUBLISHED BY SIMON & SCHUSTER

NEW YORK   LONDON   TORONTO   SYDNEY   NEW DELHI

SIMON &
SCHUSTER
CANADA

Simon & Schuster Canada
A Division of Simon & Schuster, Inc.
166 King Street East, Suite 300
Toronto, Ontario M5A 1J3

This Simon & Schuster Canada edition November 2017

SIMON & SCHUSTER CANADA and colophon are registered trademarks
of Simon & Schuster, Inc.

For information about special discounts for bulk purchases, please contact Simon & Schuster
Special Sales at 1-800-268-3216 or CustomerService@simonandschuster.ca.

Cover and illustrations by Tony Hanyk, tonyhanyk.com
Interior design by David Gee
Manufactured in the United States of America

10 9 8 7 6 5 4 3 2 1

Library and Archives Canada Cataloguing in Publication

Ingram, Jay, author
The science of why²: answers to questions about the universe, the unknown and ourselves / Jay Ingram.

Issued in print and electronic formats.
ISBN 978-1-5011-7276-2 (hardcover)—ISBN 978-1-5011-7277-9 (ebook)
            1. Science—Popular works. I. Title.
Q162.I557 2016   500   C2017-903214-3
                            C2017-903215-1

ISBN 978-1-5011-7276-2
ISBN 978-1-5011-7277-9 (ebook)

To the 2017 version of the Brady Bunch—even if they don't read it

Contents

Part 1: The Great Beyond

Part 2: The Body

Part 3: Animals

Part 4: Weird Science & Machines

The Science of Why²

Part 1
The Great Beyond

Are we alone in the universe or are aliens out there?

You can't even start to answer this question unless you believe that life—intelligent life at that—could have begun on planets other than Earth. You don't have to believe that: it's still quite possible that we are unique in the universe—that no matter how many billions of galaxies exist, containing billions of stars that have untold billions of planets orbiting them, we're the only ones. But the attitude that we're the center of everything has been eroding since the 1500s and has reduced us from being the one and only to being one of eight planets orbiting a humdrum star in one of an incalculable number of galaxies.

It's challenging to figure out whether we're alone in the universe when we don't yet have evidence of life anywhere else. But there's a way of approaching it, mostly thanks to astronomer Frank Drake, who, in 1961, invented something called the Drake equation.

The Drake Equation is a series of unknown quantities that give a sense of what we have to know before we can be confident that there are other intelligent civilizations out there. It's written like this:

$$N = R_x \cdot f_p \cdot n_e \cdot f_l \cdot f_i \cdot f_c \cdot L$$

Translated into English, the equation says that N is the number of technologically advanced civilizations out there right now that we might be able to discover. Exciting stuff! N means extraterrestrials. N means aliens!

But N is dependent on everything to the right of the equals sign. As each term is taken into account, N shrinks. That means the chances of us finding another species in the universe, then, is based on:

R_x = the total number of stars • f_p = the fraction of those stars with planets • n_e = the number of planets that are the right distance from their star to allow the existence of life • f_l = those planets that actually do support life • f_i = those where intelligent life managed to evolve • f_c = the ones that acquired advanced communications technology • and the last number, L = the number of technological civilizations that actually survive long enough for us to detect them.

When Drake came up with his equation, many of the numbers in it could only be guessed at. But since then we've managed to get a little more exact.

Science _Fiction!_ _We are most familiar with life-forms evolving on land or in water. But are there other possibilities? Two famous astronomers, Fred Hoyle and Carl Sagan, imagined weird and wild gaseous life-forms. In Hoyle's late-1950s science fiction novel_ The Black Cloud, _a giant cloud of dust and gas invades our solar system and blocks out the sun, threatening all life on earth. The cloud, more intelligent than us, lived off the energy of radiation from stars—what we call sunlight. Earth was saved when the cloud decided to move on. Carl Sagan, in a paper for NASA, put forth the idea of three kinds of giant balloon-like organisms existing in the atmosphere of Jupiter: floaters, sinkers and predators. Floaters would be kilometers in size and would survive by gathering sunlight or processing the chemicals in the atmosphere. Sinkers, like the ocean's plankton, would slowly fall through the atmosphere but could absorb other things as they fell (such as floaters), the way raindrops grow as they fall. And hunters, of course, would target other organisms to absorb._

Planets orbit stars, so, to start, we need to know how many stars there are in the universe. Our galaxy, the Milky Way, has at least 100 billion stars, and that could be roughly the same number in any galaxy. There are somewhere between 10 billion and 10 trillion galaxies, so if you multiply those numbers (using the larger estimate of galaxies), you get an incomprehensible 1,000,000,000,000,000,000,000,000. Lots of stars.

Did You Know . . . There are different estimates, but there could be as many as 60 billion habitable planets in the Milky Way galaxy.

New technologies, like the Kepler space observatory, have given us a much better idea of how many stars actually have planets around them. There are many planets out there, but we expect the ones that might hold life are those that are roughly the same size as Earth and located in what's called the "habitable" zone, where water can exist as a liquid. We earthlings assume that water, crucial to life on Earth, would be equally important elsewhere. That means that a planet can't be too close to its sun (where the heat would evaporate the water) or too distant and cold (where the water would freeze.)

We have already discovered more than 4,000 planets orbiting other stars, and it's likely that, on average, every star has at least one planet, and at least one star out of every five has an Earth-sized planet in its habitable zone. And that's not including the claim that more than 90 percent of the galaxy's planets have yet to be created. A planet's size is important, too, as it's harder for life to evolve on a giant gas planet like Saturn than it is on a rocky planet like ours.

Unfortunately, we have no idea how likely it is that a planet—even one in the habitable zone—can support life. So far we only have one example—us—in our solar system. That makes it hard to guess about elsewhere, but even if evidence of past microbial life were found on Mars, that would change the odds considerably. Scientists are hopeful that life might be common, because the chemical compounds crucial for life aren't limited to Earth at all but are found scattered all over the galaxy.

As difficult as it is to estimate how widespread life might be, what about intelligent life? Here, although it's really a guess, scientists seem comfortable with the idea that if you find ten planets with life on them, it's likely that one will have intelligent life. What's much more important is whether those intelligent species are able to become technologically adept, because only then will we be able to detect or even communicate with them.

 **Did You Know . . .** Philosopher Nick Bostrom has argued that we don't want to find other species in the universe. According to Bostrom, the rarity of intelligent life in the universe is proof that there is some event, a crucial barrier, that holds back all but a very few lucky technological civilizations, and that so far we're the only example of that.

Why is this important? If this crucial step is in our past, we're successfully through it, and the fact that we seem to be the only ones to have made it suggests that achieving technology is a very rare event. But if that barrier to becoming a fully technological, space-exploring civilization is still ahead of us—if many planets have already reached the stage we're at now, and moved on, why do we see no evidence of them?

Funnily enough, what we find on Mars is important to Bostrom's theory. He is hoping we won't find a single trace of microbial life on Mars, because that would signal that life happens often on other planets. And if that's true, it's much more likely that intelligent life, like ours, has appeared elsewhere and has been wiped out. For Bostrom, if there's no life on Mars, we can dream that we're unique. But if there is, that might suggest a bad future for our species.

That brings us to the last two numbers in the Drake equation. Detecting technologically advanced species would be awesome, but communication with them is the real goal. We have been a technological species for at the very most a few million years. (Stone tools 3.3 million years old have been found in Kenya.) And technology allowing us to communicate with distant civilizations has been around for only about a hundred years. That's not very long when you consider the lifetime of the planet—4.6 billion years—and it isn't very much time for another civilization to find us. With that timetable, aliens could have been calling us for millennia and given up long ago because we didn't answer!

 **Did You Know . . .** We have been inadvertently sending signals to extraterrestrials for longer than we've been listening for them. Before cable, TV signals used to be literally broadcast through the air. Those signals may well have been traveling through space. Just think: programs like Rod Serling's *Twilight Zone* have been traveling at near light speed since 1959, putting it somewhere between 50 and 60 light-years out there. (Would *The Twilight Zone* freak out aliens?) Unfortunately, most radio broadcasts never make it outside Earth's atmosphere.

It's clear that, despite all of our advancements, we still don't have exact numbers for all of the terms in Drake's equation, which makes it impossible to come to a conclusion about alien life. Solutions to the equation range from one civilization in our galaxy (us) possibly hosting technologically advanced life to hundreds if not thousands.

Scientists have now started varying the Drake equation to ask: How likely is it that an intelligent civilization has ever arisen in the universe? The conclusion was that unless the odds are worse than 1 in 10 billion trillion (1 in 10,000,000,000,000,000,000,000), intelligent life has to have happened. Surely the odds have to be better than that, right? Of course, we haven't heard from any of these civilizations yet, but we keep hoping they'll make contact. Maybe they're just waiting for an invitation.

Could we bring back the dinosaurs?

THERE'S A REASON THAT *Jurassic Park* was a movie and not a scientific project: making movies is easier than bringing back a dinosaur—and, even with mega-budgets, cheaper. And bringing back dinosaurs might not be the smartest thing to do.

The first challenge is to recover pristine dinosaur DNA. In Jurassic Park that DNA came from a mosquito filled with dinosaur blood that then blundered into still-liquid tree sap. That sap hardened, eventually becoming amber, handily preserving the insect for more than 60 million years.

Do you think we can make a comeback?

Worth a try.

How likely is that scenario? The closest we've come is the discovery of the remains of a 46-million-year-old blood-filled mosquito preserved in Montana shale. That's nearly 20 million years after the dinosaurs died out, so not much help in recreating *T. rex*. There are only two other known mosquito fossils old enough to have coexisted with the dinosaurs: one, from Burma, contains material that hasn't yet been analyzed; the other, from Alberta, is a male and therefore wouldn't have bitten anything. No bite, no blood; no blood, no dino DNA!

Hey, I'm trapped!
Will someone get me
outta here?

That doesn't mean we won't find the perfect specimen one day. The fact that blood is preserved in the 46-million-year-old specimen is amazing and hopeful: even the hard shells of beetles degrade over that much time. But it will likely be a flea-like insect instead. Getting our hands on the blood still leaves us far short of Jurassic Park, but there is good news: there's now evidence that we may not even have to bother with the insect.

Over the last decade several labs have found dinosaur tissue in the form of proteins preserved in dinosaur fossils themselves, a discovery that came as a huge surprise, given that fossils are, by their very nature, rock. Now, proteins are one thing, DNA is another, but the fact that there is even this much preservation suggests we're not at a dead end yet.

Let's fantasize that preserved DNA is found, it's actually dino DNA, it's intact, it can be extracted and the amounts can even be amplified so they can be worked with. Then what? Apply the techniques of cloning!

Science Fact! *Cloning has been very successful with modern mammals—remember Dolly the sheep?—but those mammals are the ideal. In Dolly's case it was easy to get a living sheep's DNA, place it into the nucleus of a sheep egg, transplant that egg into a female sheep and let pregnancy take over. But Dolly was only one success in 277 attempts!*

When it comes to dinosaur cloning, we have no DNA, no viable eggs (the ones we have are fossils) and no female. There are possible solutions. The closest living relatives of the dinosaurs are birds, so you could implant dino DNA into an ostrich egg and in turn transfer the egg into a female ostrich. (Fertilization isn't needed because the DNA already has the genetic contribution from both parents).

If all you have on hand is fragmentary dino DNA, you could add those dinosaur genes into a set of ostrich genes. That would surely be a trade-off, because the offspring would end up more ostrich than dino, but at least you'd enhance its chances of survival. Even with a complete set of dinosaur genes, the development of the embryo would be guided by sets of carefully timed inputs from the mother, so the offspring would at best be some sort of weird bird-dino hybrid. A single weird bird-dino hybrid if it indeed survived.

And that single offspring would have to eat. Birds are the descendants of therapod dinosaurs—like *T. rex* or velociraptor—who were carnivores. But a healthy existence for such creatures includes more than just meat on the hoof: gut bacteria are essential. Where would we find those? Finally, to set up a self-sustaining population, at least five thousand more animals would have to be produced and housed on a piece of land at least the area of a national park. Our ignorance of dinosaurs' ecological needs would almost certainly doom such a vast project.

If we're willing to settle for something less than a dinosaur, however, the picture gets brighter. Take the woolly mammoth. There are frozen specimens less than five thousand years old, and high-class mammoth DNA has been recovered; in fact, the complete mammoth genome has been sequenced. Second, the modern elephant, in particular the Indian species, is closely related. Genetic techniques could be used to substitute mammoth genes for Indian elephant genes, then the mammoths could be cloned.

Geneticist George Church at Harvard has already spliced mammoth genes into Indian elephant DNA—forty-plus so far. He's selected the most relevant genes for survival in cold climates, including smaller ears, hair, layers of fat, and even blood that's efficient at transporting oxygen in cold climates. That's fantastic progress, but there are still huge hurdles to be overcome.

For one thing, those forty-plus genes are only a small fraction of the genetic differences between mammoths and modern elephants. Not to mention the fact that the elephant and mammoth genes might not play nice together. For another, Church has said that he'd raise the embryo in an artificial womb: he couldn't justify experimenting with the Indian elephant, which is endangered. But that artificial womb would need to shelter an animal that could take twenty-two months to reach maturity and grow to more than 200 pounds (90.7 kilograms).

The mammoth genome chosen to represent the species is itself a challenge. It was derived from the last surviving population living on Wrangel Island in Arctic Russia, and that population had developed serious genetic flaws through inbreeding that likely contributed to its extinction.

The Russians are developing something called "Pleistocene Park" in Siberia, but even if Church is successful, there's no guarantee that reborn mammoths would have a place to live. And that's a problem that would haunt any species we bring back.

One final thought: it's likely that the money and attention given to these charismatic animals of the past—the mammoth, the dodo and the passenger pigeon—would be better spent on saving the countless species that are still alive today but are threatened with extinction.

Why is the sky blue?

Even with this straightforward question there are subtleties, but the main part of the answer is pretty clear: sunlight, which is white, doesn't travel through the earth's atmosphere unscathed. When sunlight collides with molecules of air, it's scattered in all directions. But shorter wavelengths of sunlight are more susceptible to scattering than longer ones—so light at the violet and blue end of the spectrum scatters more than red, yellow or green.

If the sun is in the east and you look at the sky in the west, the blue you see is the light that has been scattered away from the sun and then toward your eyes. However, if you look in the direction of the sun at sunrise or sunset, the sky looks red or orange. That's because the sun is near the horizon and you're looking through much more atmosphere and see only the small fraction of the sun's spectrum that's most resistant to scattering: red and orange.

So why don't we see the sky as violet? After all, violet has shorter wavelengths than blue and should be scattered more intensely by the atmosphere. There are two factors that rule that out. One is that the violet part of the sun's spectrum is less intense than the blue, so there's simply less violet light. The other is that our eyes aren't as sensitive to violet as they are to blue. But, if that's so, how can we see violet in a rainbow? Because the violet band there is separated from the other colors, whereas in a sunny sky it isn't.

There is a truly peculiar aspect to the blueness of the sky, and it's this: it's not even clear that we were always able to see the color blue. Or at least, if we could see it, we were so unaware of it that we didn't give it a name. This is, as you might guess, a controversial topic!

It starts with the ancient Greek poet Homer, the author of *The Iliad* and *The Odyssey*, two epic poems thought to have been written about 2,700 years ago. There have been many studies that have focused on these two books, but surely one of the weirdest is the one counting the number of times different colors are mentioned. The score is this: black 200, white 100, red fewer than 15, yellow and green fewer than 10 and blue: 0. Zero! The sea, according to Homer, was "wine-dark," not blue. Oxen the same. In one instance, sheep were "violet."

This Homeric study sparked interest in the subject, and subsequent surveys showed that many ancient languages also lacked words for blue and lacked names for other colors as well. In fact, there emerged an apparent cross-cultural sequence in the way color words first appeared in languages: black and white first, then red, then yellow and green, and finally blue.

So what's the explanation? Homer was a poet (except "he" was more than likely many different people); color words could simply have been the artist's choice. But the evidence that other languages only gradually incorporated color words makes it more interesting. There has also been a handful of experiments that demonstrate that different cultures appear to see, or at least to label, colors differently. In one, an African tribe, the Himba, could tell minuscule differences among shades of green that to most of us would look absolutely identical, and yet couldn't pick out the one square of blue sitting among the green.

Why the hierarchy of black and white, then red, then yellow and last—and least—blue, though? One suggestion has been that we don't need labels for colors unless we work with them. So the ancient Egyptians did have a word for blue, but they were also one of the very few ancient cultures, if not the only one, to use a blue dye. Use it, name it! This might account for the relative popularity of the word for red, a common dye and, of course, the color of blood.

Another reason offered for the rarity of words for blue is that there are few things in nature that are blue. No blue plants, very few naturally occurring blue flowers, no blue animals. Most Mediterranean or Middle Eastern peoples of ancient times weren't blue-eyed. But in North America we have birds: blue jay, blue grosbeak, blue-winged teal, bluebird. And, of course, the sky.

No one is suggesting that somehow our vision has actually changed over the last couple of thousand years. Instead, the thought is that words aren't necessarily attached to colors unless those colors acquire some importance. So how would Homer, or any other sages of the ancient world, have described the sky above them? Sure, it's sometimes leaden, sometimes even gray or black, or red and orange at sunset and dawn. But at midday?

I don't know, but I can't leave the topic without pointing out that in Samuel Butler's 1900 translation of *The Odyssey*, there's the "child of morning, rosy-fingered Dawn" (that fits), "the gray sea" (black combined with white) but also "the deep blue waves of Amphitrite." Aha! I likely haven't really discovered anything of importance, given that this is a twentieth-century translation of the ancient Greek. Butler simply may have interpreted Homer's words as signifying blue, if not doing so literally.

Roses are red,
Violets are ... blue?

Could we ever build a space elevator?

WHAT IS THE HIGHEST ELEVATOR YOU'VE EVER BEEN IN? Sixty stories? Seventy? One hundred? How about an elevator that is 12 million stories? Even on an express elevator, that would still be one long ride. But that's what it would take to create an elevator that runs from the earth's surface to the upper limits of our atmosphere—to space.

Building a space elevator isn't impossible, it's just really challenging—the equivalent to building a suspension bridge around the world. The first problem is that the elevator shaft can't be built from the ground up. There simply isn't a construction material in existence that could support something that big. You'd need a base the size of a mountain, and even with that, the whole thing would collapse under its own weight long before it reached orbit.

Going up?

But there's good news! Rather than starting from the bottom, you could assemble the elevator shaft from a satellite orbiting earth and lower it through the planet's atmosphere as it grows. Physicists love this approach: as it's being built, the tower would experience tension, not compression, and that would help prevent its collapse.

Did You Know . . . The notion of a space elevator began with the idea of so-called geostationary satellites. In 1945, science fiction author Arthur C. Clarke pointed out that if you launch a satellite into an orbit at exactly the right altitude above the equator (just under 22,369 miles, or 36,000 kilometers), it will whip around the earth at the same speed that the earth is rotating and will appear not to be moving at all when viewed from the ground.

As brilliant an idea as that was, Clarke didn't take it any further. But a Russian scientist, Yuri Artsutanov, did. Artsutanov asked: Why, if a satellite can hover over the same place on earth all the time, can't two be connected? And if an elevator shaft joined the two, then it would be possible to take an elevator to space instead of a rocket. (Although more recently the space elevator has been envisioned as a simple cable with the equivalent of gondolas running up and down.)

There is, however, one small but crucial detail standing in the way of this breakthrough. As the shaft is assembled downward, the gravitational force it feels will increase the closer it gets to the earth, dragging the satellite out of orbit. To stop that, a cable of the same mass would have to be extended above the satellite into space, exerting an equivalent force upward to counterbalance the pull of the elevator toward the earth.

This two-pronged approach would reduce some of the dangerous risks considerably, but it would bring its own challenges. For instance, by the time the elevator shaft/cable reached the earth, the outward, stabilizing extension, if it were about the same diameter, would have to be more than 62,137 miles (100,000 kilometers) long. The entire setup, upward- and downward-extending cables together, would extend about a third of the distance to the moon. A suggested solution—if it's actually fair to call it that—is to capture an asteroid and tow it into place just above the elevator, providing the necessary mass in a somewhat more compact package.

When the idea was first discussed in the 1970s, there was no material strong enough to build anything as big as a space elevator. The best suggestions were special forms of carbon, like crystals or graphite, but those were still inadequate. Luckily, the science of strong materials has come a long way since then, and nanotechnology has completely changed the picture.

The best material currently available is a peculiar form of carbon called *buckminsterfullerene*, named after inventor Buckminster Fuller because the molecules can form microgeodesic domes that resemble the ones he designed. But they also form tubes called "buckytubes." Buckytubes are incredible: they're resistant to breakage and tension (the main force that a space elevator has to fight), and a rope of buckytubes less than an inch thick would be a hundred times stronger than the same rope made of steel while weighing only one-sixth as much. But ropes of buckytubes have yet to be built. Other experimental materials are out there, but they're also a long way away from being ready for use.

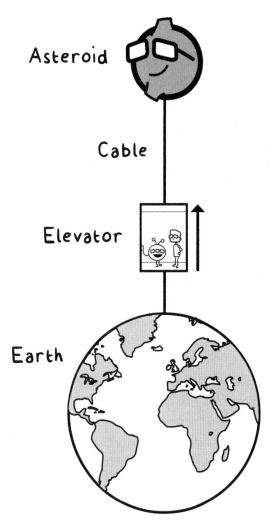

Science Fiction! *After coming up with the idea of satellites in geostationary orbit, Arthur C. Clarke proposed linking a number together into a "ring city" complete with a railroad. And in 1951 Buckminster Fuller suggested building a "halo-bridge" above the equator to which people could climb from one place on the earth, ride for a while, then descend at another.*

You might be thinking there are other issues with building a space elevator. What happens if the cable breaks? Or if a plane collides with it? Or if you get stuck 9,321 miles (15,000 kilometers) up with a bunch of people you don't like? Despite these challenges and more, though, the idea lives on.

The president of the International Space Elevator Consortium, Peter Swan, released a report in 2017 in which he defended the reasons for building a space elevator. Space elevators, he argues, will offer cheap, safe and more environmentally friendly access to space, opening up unforeseen opportunities for exploration and commerce.

Swan's vision of the elevator is a meter-wide ribbon on which small electric wheeled vehicles would climb, clinging to the ribbon by friction. He admits that it will be at least another ten years before the right carbon nanotubes or their equivalent will be available to construct the ribbon, but he remains not just optimistic but even poetic, comparing the thunderous takeoff of a rocket to the noise created by the ascending elevator as being like "dropping petals into a pond."

The space elevator might simply be an exercise in imagination—or it might turn out to be something far beyond that. But all of this fiction and possible fact shows our fascination with the idea that we might someday all have a chance to press our noses against the glass and stare down at the earth from 22,369 miles (36,000 kilometers) above.

What would happen if the moon disappeared?

THE MOON'S POCKMARKED SURFACE is evidence that over its lifetime it's been smacked by untold numbers of space rocks. Chunks of rock are hitting the moon all the time and we don't even notice, although we might have once, back in 1187. In June of that year monks in Canterbury, England, reported seeing the crescent moon split in half by fire. This might have been a piece of space rock, but it should have left a huge crater behind, and none has been found that convinces astronomers that it's the one.

Although that collision apparently didn't spark a meteor shower, more than a hundred meteorites found on the earth have come from the moon. There's no complete count of craters left by such impacts, but it would likely number in the millions.

Oww!

The moon is massive, and a similarly massive asteroid would be needed to break it apart or even nudge it out of its current orbit. As far as we know, there aren't any asteroids in our solar system like that, let alone on an orbit that would result in a collision course. So the moon will stay more or less where it is. I say "more or less" because it's slowly retreating from us—not enough that you'd notice, but each year by 1.49 inches (3.78 centimeters), about the same amount as our fingernails grow in the same time.

Why is that? Ever since the moon was created by massive collisions of space rock 4.5 billion years ago, it has been slipping away from us, because of the tides—not just the ones we're familiar with in our oceans but the reverse: tidal forces exerted on the moon by the earth. Because the earth rotates faster, it drags its tidal bulge slightly ahead of the moon, and the gravity of the mass of that bulge tugs the moon with it. That slows down the earth's rotation and at the same time speeds up the moon slightly. As that happens, the moon shifts to a higher orbit.

What if the moon actually did disappear? What difference would that make?

If you think of the earth as a spinning top, right now it's tilted about 23 degrees from vertical. It wobbles over tens of thousands of years, but not wildly. The moon's powerful gravitational influence is crucial in stabilizing that tilt. If the moon suddenly vanished, it's estimated the tilt could range from 0 degrees (straight up and down) to 85 degrees (almost lying on its side). Since the tilt of the earth is responsible for the seasons, and the slow wobble is a major factor in the periodic ice ages, a major change would provoke catastrophe.

Those dramatic tides in the Bay of Fundy and the amazing surfing in Hawaii? Pretty much over. Sure, the sun exerts some tidal forces on the earth, so tides wouldn't disappear entirely, but they'd be nothing like they are now. That would have a dramatic effect on thousands of life-forms, some of whose life cycles are precisely tuned to the lunar calendar. And I've already mentioned those millions of craters on the moon created by impacts by space rock. With no moon to intercept them, we would be exposed to the same bombardment.

And what about us? If we aren't killed by dramatic climate change or asteroids, our lives would still be altered. Stargazers irritated by moonlight washing out the rest of the universe might be pleased (although they'd miss their eclipses), but no more moonlit nights for the rest of us. The lack of a moon would also disappoint romantics and writers of cheesy songs. No more conspiracy theories about whether astronauts really went to the moon or not. No more howling at the moon, nothing for the cow to jump over and certainly no more strange human behavior during the full moon.

But wait! There is no strange behavior during the full moon, despite what you've heard. No more crime, no more hospital admissions, no higher rate of births, no more suicides or homicides—nothing. Rumors persist, but studies reveal the truth: the full moon causes no apparent change in human behavior.

Why is the night sky dark?

THAT SEEMS LIKE AN EASY QUESTION. Even on the most spectacular nights for stargazing, most of the sky is dark. A star here, a star there, but pinpricks—not enough to light up the night. It's so dark that intergalactic space is more than a million times dimmer than the light you're reading by.

At first glance, it makes sense that it would be dark in space. But something's missing. Stars give off light. And there are billions of them. The number of stars in our Milky Way Galaxy alone is probably 100 billion. And that's just our galaxy. There are hundreds of billions more galaxies out there, as the Hubble Space Telescope has shown us. That guarantees that wherever you look in the night sky, you can be sure that you're staring at a wall of stars, or, if you prefer, a wall of galaxies.

The physics says it doesn't matter whether all the stars are clumped into galaxies or are off by themselves; it should still be bright out there. The physics also concludes that any patch of the night sky, no matter where you look, should be about as bright as the surface of the sun. So there's a huge puzzle here. It's called Olbers' paradox. There are two theories that carry some weight when astronomers argue about Olbers' paradox.

One theory attributes the darkness to the expansion of the universe, which has been going on since the big bang. And it isn't like everything is flying apart as it does when there's been an explosion in the kitchen. Instead, space itself is expanding.

Because of that expansion, the space between the galaxies is increasing, with the result that galaxies are receding from us. As that happens, the light from the galaxies undergoes a physical shift that we are familiar with from sound: the Doppler effect. Listen to the whistle, siren or sounds of the engines as a train or ambulance nears. As it approaches, the sound is high-pitched; as it recedes, the sound is low-pitched. It happens because sound waves are either pushed together (higher pitch) as the vehicle approaches, or stretched out (lower pitch) after it's passed.

Looks like the universe isn't the only thing expanding.

The same happens to light. When a light source is receding at rapid speed, the entire spectrum of visible light is stretched à la the ambulance in the previous example. And when light waves are stretched enough, they exit the visible spectrum to become infrared light, microwaves or even radio waves, all of which are invisible to us. Even so, this effect dims the light only by about a factor of two, not enough to make the night sky as dark as it is. Also, some scientists have pointed out that shifting of light due to high speeds should also happen at the other end of the spectrum, where invisible waves of ultraviolet light should stretch and become more visible rather than lost. So the mystery remains.

The other major factor that relates to darkness in space is the age of the universe: 13.82 billion years. No galaxies are older. In fact, the first stars began to shine about 200 million years later than that date and the earliest galaxies somewhat after that. That means that no galaxy that we can see can be any farther away than 13.82 light-years: that's the maximum distance that light from that galaxy could have traveled in that time. Because the speed of light is finite, there will be galaxies farther away whose light hasn't had time since their births to reach us. It's also true that galaxies die and wink out, and that new galaxies are coming to life all the time, but the continuing and even accelerating expansion of the universe guarantees that many will always be out of range. Imagine there's an opaque dome roughly 13 billion light-years away—we will never see anything on the other side of it.

The paradox of "so many stars, so little light" has attracted big thinkers over the years. Famous comet man Edmund Halley wrote about it; eighty years later, Wilhelm Olbers donated his name to it and came up with a theory that light was being intercepted by material between the stars. Unfortunately, he failed to give credit to the man who anticipated his ideas, but they were wrong anyway, because any material that did intercept starlight would heat up and eventually become hot enough to radiate light itself.

The most unexpected contribution, though, came from Edgar Allan Poe, the poet, short story writer and, apparently, cosmologist. In "Eureka: A Prose Poem," published in 1848, he anticipated the importance of the age of galaxies. He argued that the only way to explain the darkness was to imagine that the invisible background was so immense that no ray from it had been able to reach us.

Poe's insight was pretty amazing, considering how little was actually known about the universe at that time. As it turns out, he was right: the sky, as seen from earth, will always be dark because most of the light that's out there is too far away to reach us.

What's dangerous about the Bermuda Triangle?

THE BERMUDA TRIANGLE HAS A BAD REPUTATION. A huge expanse covering more than 500,000 square kilometers of ocean, it is associated with more than a hundred unexplained disasters. The points of the triangle are anchored by Florida, Puerto Rico and, of course, Bermuda.

So is there something truly dangerous about the Bermuda Triangle? And, if so, what is it? Some theories argue that "crystal energy" accounts for the strange disappearance of planes and boats in the triangle. Others say that alien interference in the region is to blame. Still others point the finger at spirits. Or a giant squid. Or an interdimensional doorway. Before you go too far down the supernatural spiral (taking your spirits and giant squids with you), let's check out the scientific evidence.

Accounts of ships and planes disappearing in the Triangle have been documented for well over a hundred years. The most famous of these disappearances occurred in 1945 when five TBM Avenger torpedo bombers with the U.S. Air Force (later named the Lost Patrol) disappeared. The planes took off from Fort Lauderdale just after 2 p.m. on December 5 on a routine practice run. The plan was to fly due east over the Atlantic Ocean to engage in a brief practice bombing, then fly farther east, then north and finally west again to return to base. The weather was excellent when the fleet took off, but the planes never made it back to base. No trace of any of the aircraft was ever found—no wreckage, no bodies. Fourteen lives were lost, and the mystery remains unsolved.

The only evidence we have are the radio reports from the pilots and they suggest the pilots were confused as to where they were:

> *"I don't know where we are. We must have got lost after that last turn."*

> Flight leader: *"Both of my compasses are out and I am trying to find Fort Lauderdale, Florida. I am over land but it's broken. I am sure I'm in the Keys but I don't know how far down and I don't know how to get to Fort Lauderdale."*

> Flight leader: *"Change course to 090 degrees [due east] for 10 minutes."*

> *"Dammit, if we could just fly west we would get home; head west, dammit."*

> Flight leader: *"Holding 270, we didn't fly far enough east, we may as well just turn around and fly east again."*

> Flight leader: *"All planes close up tight . . . we'll have to ditch unless landfall . . . when the first plane drops below 10 gallons, we all go down together."*

These descriptions suggest that the five pilots of Flight 19, including the flight leader(!), became confused about their location. They likely flew in more than one wrong direction as they tried to find their way back to Florida, finally running out of fuel and crashing into the ocean. True, their compasses had stopped working and the weather turned stormy. But still, there is no clear explanation for why all five would have gotten so lost.

The disappearance of those five planes was strange enough, but another thirteen people died right after when a different aircraft—one of a pair sent out to search for the missing planes—was also lost. The rescue plane, a PBM-5, radioed back to base three minutes after takeoff, and that was the last it was heard from. Twenty minutes later a ship in the area, the SS *Gaines Mills*, reported seeing an explosion and "flames 100 feet high." Shortly after, the ship sailed through a slick of oil and aircraft fuel. The downed rescue plane was reputed to have had recurring problems with fuel leaks. Even with precise details about the location of the explosion, not a single piece of that aircraft was ever found.

Science Fiction! *The Bermuda Triangle is the final resting place for the city of Atlantis . . . or so say Paul Weinzweig and Pauline Zalitzki. The two scientists claim to have captured images of sphinxes and pyramids of the Lost City on the ocean floor in the Bermuda Triangle. There's just one problem: given the location of these findings, Atlantis could only ever have existed hundreds of meters below sea level, which would have made it a very wet city. Weinzweig has acknowledged that the pyramidal shapes he found on the ocean floor might be natural formations, not from Atlantis at all.*

The USS *Cyclops*, a ship that disappeared without trace in 1918 with more than three hundred people on board, is another well-known disappearance in the Triangle. The ship was likely vulnerable to stormy weather because it was carrying about eleven thousand tons of manganese ore, making it tons over capacity and seriously overweight. Two of its sister ships later sank because of suspected structural flaws.

The latest attempt to explain the supposed "weirdness" of the many disappearances in the Bermuda Triangle brings more science into it. The Science Channel in the USA promoted a theory that unusual hexagonal cloud formations seen in satellite images over the Triangle might be connected to the disasters there. According to the report, the hexagonal clouds generate microbursts, downward "air bombs" that travel at speeds of up to 186 miles (300 kilometers) per hour. These would obviously be hazardous to boats and planes, but so far this theory has not been proven.

The bottom line is that while there are no definitive explanations for the disappearances in the Bermuda Triangle, in most cases, there's some rational evidence to suggest mechanical failures and weather events as opposed to squids, crystals and spirits. In the end, it seems that the most dangerous thing about the Bermuda Triangle is simply believing in it.

What are near-death experiences?

IT'S EASY TO DESCRIBE what the typical near-death experience (NDE) is like, but so far it's been impossible to explain exactly what it is.

It's not uncommon that someone who is either severely ill or gravely injured comes to a point where they're close to dying: their heart has stopped, and electrical activity in the person's brain has flatlined. If that situation were to persist, it would be death. But there are many who have pulled back from that brink and continued living a normal life.

Of those, a small number report having had a near-death experience. That could involve an out-of-the-body experience: hovering over the operating table looking down on their unconscious body, or traveling down a tunnel with a

I'm having an out-of-body experience.

Me too.

light at the end of it, meeting deceased family and friends, feeling at one with the universe or encountering some sort of spiritual being. There have also been unpleasant NDEs—the feeling of complete nothingness or being in a place peopled by demons and threatening animals—but those are much rarer than the positive ones.

What could be happening to create such vivid images in a person who's near death? It comes down to this: if a person has an NDE when their brain activity is zero, it suggests that mental activity can happen independently of the brain. That idea is embraced by many—it's a standard part of the belief in a "soul"—but scientists are convinced that our mental life is generated by the brain and exists nowhere else.

In 2001 a Dutch scientific research team reported in the medical journal the *Lancet* their studies of NDEs experienced by cardiac-arrest patients. They identified a short period of time during which the heart had stopped and the brain presumably flatlined as the only possible time when these patients could have had their experiences. The researchers asked: How could a clear consciousness outside one's body be experienced at the moment that the brain no longer functioned?

I think,
therefore you are.

Fewer than 10 percent of the patients they studied had a substantial NDE, but the details reported by those who did were eerie: one man remembered specific things doctors were doing around him, including removing his dentures. He even sensed the pessimism of the medical team.

Reaction to the Dutch team's findings from other scientists was predictable. An editorial in the same issue of the *Lancet* argued that it was difficult to prove that the patients' NDEs happened exactly as their brains were flatlined as opposed to just before or after. It also raised questions about the veracity of the reports, because several people who originally said they hadn't had an NDE changed their minds when asked again two years later. The Dutch team argued that because the majority of their patients hadn't experienced an NDE, the standard scientific explanations of the brain responding to high levels of carbon dioxide or low levels of oxygen didn't make sense. Otherwise, virtually all patients would have had those experiences.

The doubts about the 2001 study focused the challenge sharply: When there is sure evidence that the brain has flatlined, is there ever any mental activity? And how can the stories of people claiming to have had NDEs be confirmed?

Did You Know . . . In 1977 a woman named Maria in a Seattle hospital suffered cardiac arrest. She recovered, and the next day she reported having seen herself above the operating table, then looking outside and seeing a distinctive tennis shoe on the window ledge. The shoe was located and the details fit Maria's description. This seemed to suggest an NDE that involved an out-of-the-body experience. But later two skeptics returned to the hospital and placed a shoe on the same ledge. They noted two things: First, the shoe was easily seen from the ground and could well have been mentioned in Maria's presence. But even more damning was that every detail of the shoe could be seen from inside the hospital room, even from the bed.

A recent study took advantage of the fact that many NDEs include an out-of-the-body experience, usually of the patients looking down on themselves and the medical teams. For these to happen at all is almost unbelievable, let alone for them to happen during cardiac arrest! In this case the researchers installed shelves in the operating rooms where emergency procedures were likely to happen, then put objects on those shelves that could be seen only by someone floating in midair.

In more than 2,000 cases of hospital admission for cardiac arrest, there were only 330 survivors and only a handful of interesting cases of NDEs. In the end, only one patient was able to describe verifiable events while he was in cardiac arrest. Sadly, there were zero instances of patients describing anything hidden on the shelves.

There are occasionally small pieces of research that show that there's more to discover. One Canadian study in early 2017 found that of four patients who had life support withdrawn, one continued to show peculiar brain wave activity for ten minutes after the heart stopped. It was only a case of one, but it was completely unexpected. And another recent study in rats showed intense activity in the brain immediately after cardiac arrest—a burst that resembled—and even exceeded—the brain activity seen in awake rats. That also came as a surprise. Was this activity due to a shock to the brain, or was it rodent enlightenment?

The frustrating thing about the divide between believers in NDEs and nonbelievers is that the controversy sits right at the heart of one of science's most challenging puzzles: consciousness. What goes on in our brains to generate the thoughts, dreams, ideas and images that we experience when we're conscious? Is the brain necessary for those things? Many who believe in NDEs think not; scientists disagree.

I stink, therefore I am.

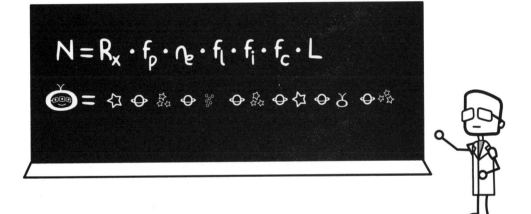

History Mystery

How did the rainbow come to have seven colors?

WE OWE THE SEVEN COLORS of the rainbow to Sir Isaac Newton, one of the most brilliant—and difficult—scientists who ever lived. He is often caricatured as a plump little man getting hit in the head with an apple, but that doesn't do Newton justice: he was a one-of-a-kind scientist. In just a year and a half in his early twenties, Newton invented calculus and came up with theories to explain both gravity and light. It became known as Newton's *annus mirabilis*, a somewhat inaccurate title, as it was actually an *annus* and a half.

We think of Newton as a great scientist, but he also dabbled in the occult. He was an alchemist and a theologian, writing more than 2 million words on those subjects—twice as much as he wrote on science.

Because of these writings, John Maynard Keynes refused to view Newton as the first giant of science, calling him instead "the last of the magicians."

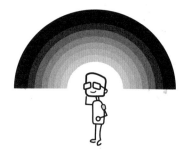

There's certainly magic in the colors of the rainbow. We're so familiar with them that we can't really imagine how shocking it was when Newton first proposed that the rainbow was actually sunlight that had been broken up into its colors.

Did You Know . . . There are wavelengths outside of the visible spectrum that are either too long or too short for our retinas to pick up. But it's easy to imagine the rainbow continuing beyond its ends: violet fading to ultraviolet; red giving way to infrared . . .

For many years Newton claimed that sunlight was made of only five colors: red, yellow, green, blue and red. In at least twenty-seven different lectures, Newton defended that number. Then, suddenly, he announced that both orange and indigo should be added to the rainbow—indigo slotted in between blue and violet; orange between red and yellow. What had changed to prompt Newton to make this addition?

Newton had already been able to separate sunlight into its constituent colors (or wavelengths) and placed them in the familiar sequence, from red to violet, measuring their distances on the spectrum relative to each other. Even so, he was always careful to explain that, while there were distinct colors, each gradually merged with the next with no sharp divisions. On the spectrum, the five fundamental colors—red, yellow, green, blue and red—stood out, but there were places in between where you might or might not be able to discern other colors, rather than just a gradual mix. Newton took advantage of this to introduce his two new colors.

Some have suggested that Newton wanted to establish seven colors because the number seven had mystical properties. As "the last of the magicians," he embraced the number seven: it described not only the solar system (at least at the time) but also the common metals. Others think he simply used artists' color "circles" to map out his own ideas about color, and among them would have been a few with the visible spectrum divided into seven. Some suggest the painter's imprint is there in Newton's own color circles: the locations of some of the colors make sense to a painter, not a scientist.

But my favorite suggestion is that he wanted to connect the colors in sunlight with music, perhaps because both had mathematical underpinnings. Newton had already written about the math connection between music and gravity. The musical scale in Newton's time was symmetrical, consisting of five notes a whole tone apart and two extras only half a tone from their immediate neighbors—seven notes total. The theory argues that Newton recognized that his scale of light could match the musical scale perfectly if he inserted

orange and indigo where he did. It made his theory of light more comprehensible by likening it to something everyone was familiar with, but it was, at best, an approximation, because the musical notes are determined by one kind of mathematical relationship, and rainbow colors by another.

Exactly why Newton wanted to make the connection to music isn't clear, although at one point he admitted that he wasn't particularly good at discerning the border between any rainbow color and the next. The musical scale might have helped make those judgments. Regardless, it could be that Newton used music to help create the rainbow spectrum we know today.

Did You Know . . . One of the most famous depictions of Newton's rainbow is the album cover of Pink Floyd's *Dark Side of the Moon*. Most people don't realize that the rainbow on the cover has only six colors, not the standard seven, or even the five that Newton played around with. And the missing color? Looks like indigo to me.

Part 2
The Body

Why do I get hiccups and how can I make mine stop?

LET'S TAKE THESE TWO QUESTIONS one at a time. In fact, the first question—why we get hiccups—is itself two questions: what happens in a hiccup, and why are our bodies are prone to them?

What happens in the body is a little complicated: it involves the brain and muscle groups in the chest, in particular the diaphragm. The diaphragm lies just above the stomach, and its movements enable breathing: when the diaphragm relaxes, it rises and air is forced out of the lungs; when it contracts, it moves downward and air is sucked into the lungs. When we hiccup, a part of the brain called the medulla signals the diaphragm to drop immediately. Normally, that would mean that you'd inhale a big gulp of air, but precisely thirty-five-thousandths of a second after that big gulp starts, the glottis, the space between the vocal cords, closes abruptly, shutting

Hic!
Hic!

down the incoming air. That's a hiccup, and the sound of it mirrors exactly what happens: a very, very brief inrush of air that's cut off almost immediately.

Hiccups are much more common in infants than in adults, and even more so in fetuses. In fact, fetuses just eight weeks old have been seen hiccupping in ultrasounds, for several minutes at a time, even though their diaphragms aren't yet fully formed. As we get older, we hiccup less and less.

Science _Fact!_ Hiccupping seems to be confined to mammals. It has been seen in rats, cats, rabbits, horses and dogs, but never in any reptile, bird or amphibian.

There are two main causes of hiccups. One is an immediate reaction to a stimulus. There's a range of things that can prompt a hiccup, including too much alcohol, too much food (especially if it's really spicy), bubbly beverages and prolonged hysterical laughter. Most of these generate hiccups that might last for a few minutes or even an hour or two before they subside. But hiccups may also be triggered by a wide variety of medical conditions, including acid reflux, ulcers, a skull fracture, various infections, multiple sclerosis and blood clots in the brain. In one bizarre case, hiccups were set off by a hair inside a person's ear canal tickling their eardrum. While it may seem odd that such a diverse set of conditions can cause hiccups, when you consider that the brain, nerves traveling between the brain and the diaphragm and many other muscles are all involved in a single hiccup, it makes more sense.

Long-lasting cases of hiccups are more common than you might think. One study that covered the years 1935 to 1963 found 220 cases of hiccups that lasted at least two days, with most continuing for more than two months. Men were nine times as likely as woman to be plagued with continuous hiccupping. Most persistent hiccuppers had periods of several days when the hiccups happened all the time, followed by days when there were none.

 **Did You Know . . .** The world record holder for the longest continuous bout of hiccups is the late Charles Osborne, who started hiccupping in 1922 when he was twenty-eight years old and continued until 1990—sixty-eight years of hiccups! You might think that his hiccupping would have completely derailed his life, but he raised eight children and ran successful businesses. The sad irony is that after hiccupping for sixty-eight years, he lived only a few months after they stopped.

Why does the body have a mechanism for hiccups in the first place? Unlike vomiting, gagging and coughing, hiccups are not very effective at expelling noxious substances from the body or clearing airways. The facts that hiccups are most common early in life and that only mammals hiccup prompted Dr. Daniel Howes at Queen's University in Kingston, Ontario, to suggest that the hiccup is crucial for mammalian infants. Howes argues that nursing infants can swallow a lot of air as they nurse, and that air can get in the way of incoming milk. By creating a bit of a vacuum in the chest, the hiccup draws that air up the esophagus, where it can then be burped out, leaving room for more milk.

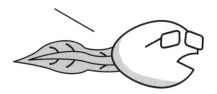

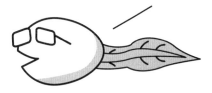

Another provocative idea is that hiccupping is a hangover from a very distant time when tadpoles (or their ancestral equivalent) were adapting to breathing, either through gills or lungs. Switching from one to the other involves closing off the flow of air to the lungs, while opening up the flow of water through the gills, not unlike expanding the chest and closing the glottis. We have many neural programs that go back millions of years, why couldn't hiccupping be one of them?

Given the complexity of a hiccup, it's fitting that the list of hopeful remedies is long. Drug treatments are reserved for the most serious cases, but there are plenty of folk remedies, some of which are more scientific than others.

 TRY THIS AT HOME! Next time you have the hiccups, try swallowing ten times in a row, then holding your breath. This is my favorite, and the science behind this folk remedy makes sense: when you hold your breath, you raise the level of carbon dioxide in your blood. Exactly why carbon dioxide works isn't clear, but somehow it calms the diaphragm, neurologically speaking, making it less likely to trigger a bout of hiccups. The same thing happens if you breathe into a paper bag.

 DON'T TRY THIS AT HOME! My parents taught me that drinking a glass of water from the opposite lip of the glass cures the hiccups. In other words, tilt the glass away from you while you drink. The only problem: this usually leads only to more hiccupping and a lot of spilling.

The much-quoted *New England Journal of Medicine* published data suggesting that a spoonful of raw sugar would cure the hiccups. In one study, that worked nineteen out of twenty times, good enough to rank near the top of the methods in the study. In a 1999 article in the *Canadian Family Physician*, Ronald Goldstein claimed that having a hiccupper completely plug their ears and then drink a large glass of water through a straw would do it. Unfortunately, he quoted no statistics to support his theory.

In fact, a lot of studies of hiccup remedies are thin on numbers. In a research paper published in 2000, sex was suggested as the perfect remedy for the hiccups. But the study referenced only one case where this worked: a male who had been hiccupping continuously for four days. Massaging the far back of the throat with a Q-tip for a minute is also said to have worked—but only once.

The most surprising remedy of all is surely digital rectal massage—meaning massage with the fingers. Even though there are three different reports published on digital rectal massage as a cure for hiccups, there are still only a handful of patients who've tried it. I suppose that's not surprising. Other than saying that there are a lot of nerves fanning out from the rectum and making connections elsewhere in the body, we are left guessing as to how this technique would alleviate the hiccups. And so far we are having trouble amassing a crew of volunteers for this research.

Why can't I tickle myself?

AT FIRST GLANCE, this question does not seem to be of profound scientific importance, but the fact that scientists from Charles Darwin on have been fascinated by tickling suggests otherwise.

There are two kinds of tickling. One is the result of dragging something like a feather or cotton swab very lightly across the skin. This sensation is annoying more than tickly and triggers not laughter but a sudden, even violent withdrawal. It feels like an insect is crawling across the skin.

It's much firmer tickling that makes us laugh, often uncontrollably. Even though we laugh, it's debatable whether we're actually enjoying ourselves, since we're also struggling to get away. So the question leads to another question: Why do we laugh when we're tickled?

Darwin argued that the commonality between laughter induced by tickling and by a joke is the element of surprise. You can't tickle yourself, and you can't tell yourself a joke: neither is surprising. In the early 1970s a British research group created an ingenious tickling box to test the idea of surprise. Volunteers placed their feet in it and a plastic pointer tickled them.

The key feature of the tickling box was that the pointer could be controlled by the person being tickled or by someone else. Thirty undergraduate students took the tickle test. The results? The test subjects were much more ticklish when someone else was controlling the pointer. And when blindfolded students were tickled by a machine rather than a person, they laughed just as hard, proving that the tickler doesn't have to be an actual person.

A variant of the experiment used a feather instead of a pointer. If someone other than the ticklee held the feather, the tickling sensation was extreme. But if the tickler put the feather in the ticklee's hand but held the hand and controlled the feather's movement, the ticklishness was diminished. It would seem that, again, the element of surprise made the sensation stronger, but if the tickling could be predicted, it didn't have the same effect.

The experiment was fine-tuned to discover if the timing of the tickling movements was important. An apparatus was set up that allowed a person to tickle themselves using a robotic hand. If there was no delay between the person's hand movement and the resulting tickle, the ticklee didn't laugh. If there was a delay between the hand movement and tickle, the ticklee did laugh. The longer the delay, the more ticklish the touch, apparently because it felt more like someone else was doing it.

Then a different British team used brain imaging to test tickling further. The team was able to show that when a person engages in self-tickling, the part of the brain called the cerebellum is activated. This isn't surprising: the cerebellum is a large area at the back of the brain that plays a major role in coordinating movements. You work your cerebellum when you play the piano or tighten a screw. The researchers suggested that, in addition to coordinating the tickling movement, the cerebellum let other parts of the brain know that tickling was about to happen.

But why do we often laugh hardest when we know the tickling is coming? A team at the Karolinska Institute in Stockholm answered that question by comparing brain activity in people who were actually being tickled with those who were about to be tickled. The startling result was that there seemed to be no significant difference. The scientists interpreted this to mean that the brain was preparing itself for action. Most of time when you're threatened with being tickled, you will be tickled. This particular situation isn't life-threatening—although it can sure feel like it—but in many other cases the threat of physical action would be. A brain that anticipated coming events would be better prepared to act and survive.

There is one intriguing exception to the rule that you can't tickle yourself: schizophrenics. Schizophrenics often have trouble distinguishing their own actions from the actions of others. Hallucinations are a good example of this: hearing a voice they attribute to others, or feeling their lives are being controlled by others. So, just as schizophrenics sometimes think the voices they hear belong to others, they think their own tickle belongs to someone else. In fact, it's not only people with the full-blown illness but even those who share some tendencies with schizophrenics, like delusions and odd behaviors and beliefs.

It might even be possible to add dreams to that list of tendencies, because they're full of hallucinations—made-up dramas happening at a time when we're taking in little sensory information. One experiment indeed demonstrated that some subjects were more sensitive to self-tickling immediately upon waking out of a dream. Unfortunately, and somewhat mysteriously, they were all women; the handful of men in the study didn't exhibit that tendency. But since the study was very small, it wasn't conclusive.

Trying to understand why you can't tickle yourself may seem unimportant, even frivolous, compared to many other scientific projects, but it actually is linked to mental health and closely tied to our ability to distinguish ourselves from others.

What is a hangover and how is one cured?

HUMANS FIRST STARTED DRINKING about eight thousand years ago, and the first hangover was probably the day after. Yet, surprisingly, there are still questions surrounding what causes hangovers and how they might be prevented.

People are going to drink, hangovers or not. (One estimate is that 75 percent of people who drink have had one.) One scientific study of hangover remedies in the *British Medical Journal* admitted that "no conclusive evidence shows that hangover effectively deters alcohol consumption."

While each person's hangover is uniquely unpleasant, there is a generally accepted (and very long!) list of symptoms, including: fatigue, headache, drowsiness, dry mouth, dizziness, nausea, sweating, anxiety and a variety of mental effects, like the inability to concentrate and to remember.

One striking fact: the hangover begins as the blood alcohol level declines, then peaks when alcohol disappears from the bloodstream and continues for as long as another twenty-four hours. Why is that? Two suspects are the metabolic products of alcohol and the other chemicals present in the drinks consumed.

Alcohol is processed into a chemical called acetaldehyde, which degrades first into acetate and finally into carbon dioxide and water. High levels of acetaldehyde are known to provoke facial flushing, increased heart rate, lower blood pressure, dry mouth, nausea and headache, all of which fit nicely with hangovers. However, acetaldehyde is broken down quickly in the body, suggesting its role might not be that significant.

Experiments with rats suggest that acetate might be more important. The rats ingested pure ethanol and were tested for their pain threshold by probing around their eyes with short nylon filaments. Sure enough, they retreated from the filament at about the same time as the amount of alcohol in their systems was zero—in other words, just as the human hangover kicks in. They didn't react to acetaldehyde that way. The good news? Caffeine seemed to relieve the pain—interesting, since coffee was touted for its hangover relief more than a hundred years ago.

The other culprits in hangovers are substances called congeners, chemicals in various types of alcohol that are created in the fermentation or distillation process. There's a ranking system based on resulting hangovers: gin and vodka have the fewest of these additional chemicals, while red wine, bourbon and especially brandy have the most.

Of all human behaviors, drinking might be the one that's most resistant to that ultimate of study designs: the randomized double-blind trial, in which participants receive a placebo or alcohol, but neither they nor the experimenters know which they're getting. We've all heard stories of people at a party who are fooled into thinking they're drinking and who then act drunk, but in a lab setting? Do participants consuming a placebo not know that's what they're drinking? Can the experimenters not tell? The difficulty of establishing what exactly provokes symptoms stands in the way of coming up with a cure.

While a Google search for "hangover cures" or "hangover remedies" generates hundreds of thousands of hits, a quick glance reveals that almost none have been scientifically evaluated. No surprise.

One scientific review contrasted a short list of hangover remedies touted on the Internet (including aspirin, fresh air, honey, cabbage, and at least three versions of drinking more alcohol) with a much smaller set of actual studies and concluded that even the actual published-in-the-scientific-literature studies didn't amount to much. Those that hinted at some effective hang-over treatment, like prickly pear, artichoke or dry yeast, either didn't include enough subjects or had some other constraint that required the study be repeated. The authors concluded there was no evidence for any effective intervention.

So I'll end with another quote from the same article in the *British Medical Journal*: "The most effective way to avoid the symptoms of alcohol induced hangover is to practice abstinence or moderation."

What is the Law of Urination?

WHILE THE LAW OF URINATION doesn't quite carry the impact of the law of gravity, it does affect our everyday lives, much as gravity does, and in fact is dependent on gravity.

The Law of Urination is: all large animals (anything over about seven pounds, or three kilograms) take about the same time to pee. Of course no law of any kind comes down to a single exact number. There are always outliers, and so the law is more precisely worded: large animals empty their bladders over a nearly constant duration of 21 seconds plus/minus 13 seconds. That is, the slowest take a little more than half a minute, the fastest a mere 8 seconds.

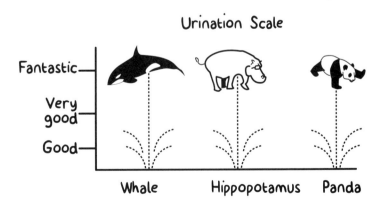

Urination Scale

Fantastic

Very good

Good

Whale   Hippopotamus   Panda

Did You Know . . . Small animals—say, the size of a rat—cannot generate a stream of urine and are limited to drops.

You would be missing some totally fascinating details if you just left it at that, because even though the Law of Urination is the subject of only one scientific paper, that paper has much, much more of interest in it than just a couple of numbers. For one of the first times in the history of science, YouTube videos provided some of the evidence. Twenty-eight videos from the internet were combined with sixteen that David Hu and others at the Georgia Institute of Technology shot at Zoo Atlanta and other locations. These videos allowed them to time urination over a range of animals that differed in size by a factor of more than 200,000, from 0.066 pounds (0.03 kilograms) to 17,637 pounds (8,000 kilograms). Given that enormous size range, how is it possible that there's a fairly narrow time range for peeing?

Plumbing is a big part of the answer. For all animals, the urethra, the tube that connects the bladder to the outside world, is twenty times longer than it is wide. That compares to the dimensions of a coffee stir stick. A rat's is like a tiny piece of string, an elephant's more like the drain under the sink. But the length-to-width ratio is constant over that fantastic size range.

The bladder is also crucial. It's a muscular sac that exerts pressure to expel urine. That pressure is about the same as exerted by a column of water 19.3 inches (49 centimeters) high. The size of the bladder is consistent as well, with bigger animals having the same volume of bladder, proportionally.

The pressure exerted by the bladder and gravity are the two factors that most affect the flow of urine. Animals with big bladders also have long urethras. The farther along the urethra, the greater the pressure because of the weight of the urine above, just as the pressure increases toward the bottom of a swimming pool. So bigger animals achieve a faster flow of urine, which allows them to drain quickly enough to keep pace with much smaller creatures that have less urine.

In their analysis, the authors of the paper made two assumptions: one, that animals pee when their bladders are full; and, two, that animals are directing their urine downward, thus harnessing the full force of gravity. These assumptions may hold most of the time, but there are notable exceptions. For one, anyone walking a dog knows that either they can somehow regenerate urine at a fantastic clip, thus allowing them to pee every minute and a half, or, more likely, they hold back, because they're marking territory, and one wouldn't want to run out of marking fluid at a crucial time!

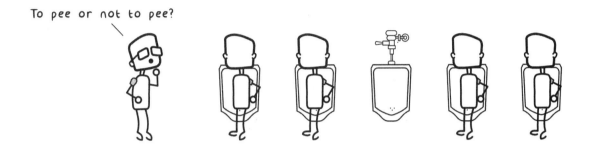

To pee or not to pee?

As for gravity, some animals go to great lengths to defy the force. Male pandas, for instance, as a measure of superiority, do their best to deposit their urine as high up a tree as possible. They actually do the panda equivalent of handstands while peeing, and, by doing so, they lose most of the gravitational boost they otherwise would have if they'd peed "normally." But that's really what dominance is all about, isn't it? A truly magnificent specimen of a panda doesn't even need gravity.

Hey! Urine my tree!

I sure am.

Science _Fact_ or _Fiction_! There's a legend (unverified) that in 1646 the scientist Blaise Pascal invented something called Pascal's barrel. He inserted a thirty-three-foot (ten-meter-long) tube into a barrel full of water. He then filled the tube with water as well. The pressure exerted by the water in the tube broke the barrel apart. This is the same mechanism that allows very large animals to pee at the same rate as small ones.

Apparently heartened by the success of this study the same scientists went on to measure the time it takes a variety of animals to poop, with similar results. No matter what the size of the animal, from cats to elephants, each requires 12 seconds plus or minus 7 seconds to poop. The study was published in the journal Soft Matter. Yes.

Why can't I remember anything that happened before I was two years old?

You may already be thinking, "But wait! I distinctly remember my first birthday!" or "My first trip to Niagara Falls was when I was one and a half!" But those memories could easily have been created by seeing photographs or hearing events described by others, over and over, not necessarily by remembering the actual experience. Still, there are some exceptional people who really do remember things from a very young age. But no more than 1 to 2 percent of the population seem to be able to do that.

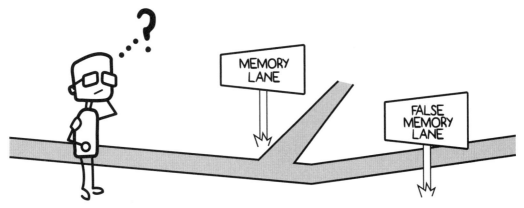

For the vast majority of people, first memories can be recalled from the age of about three to three and a half years. A child's brain is rapidly developing before that, and somehow, during that crucial process, stored memories seem to mysteriously vanish.

Researchers have tried to prove that children under the age of two can actually store memories, but because very young children lack a good command of language, proof is difficult. But New Zealand psychologists Harlene Hayne and Gabrielle Simcock found a way around this problem, by building something called the Magic Shrinking Machine.

The Magic Shrinking Machine was used with children ranging in age from just over two years old to just over three. Each child learned to start the machine by pulling a lever that turned on lights. One of the experimenters then put a large toy into the machine, making it "disappear," then the same experimenter turned a handle to produce a set of sounds. Finally, the child was shown how to retrieve the toy from the machine. Lo and behold, when the child retrieved the toy, it had shrunk (or at least appeared to have shrunk)! The child repeated the actions seven times, each time with a different toy that magically shrunk. By the end of the experiment, the children were able to repeat the entire sequence by themselves. This proved they clearly remembered what to do.

Six months later, then a year later, the children were tested to see if they could remember the Magic Shrinking Machine and how it operated. They were tested using two different methods. One was verbal: they were asked questions about the machine, like "Last time I visited you, we played a really exciting game! Tell me everything that you can remember about the game. What were the names of the toys? And how did we make the Magic Shrinking Machine work?"

Then the experimenters tried to coax nonverbal memories by showing the children pictures of the machine, the toys, and the bag the toys had been carried in. They even showed the children the actual machine to see if they could operate it. They found that the younger the child had been at the time of the experiment, the less they remembered, and the more time had passed, the worse the memories were. They also found that if a child lacked the vocabulary to describe the machine in the past, they couldn't describe it later—even if they'd acquired the appropriate vocabulary in the meantime, and even if the nonverbal tests showed that they remembered the machine in some detail. This suggests that our ability to talk about our early memories is limited by our ability to process language.

How to jog your memory

Six years later the children were tested yet again. Many remembered the machine, even those who were only two years old at the time they'd first been introduced to it. That finding was surprising. It showed that the language barrier wasn't as impenetrable as once thought. But it was still significant for a lot of test subjects.

A number of theories have been proposed to explain what is necessary for us to form crystal-clear memories. One is that we can't do that before we realize we have an identity. That happens sometime around the age of two—roughly the same time that we start speaking and understanding words. It's not yet understood how those two events relate, but there is evidence that the kinds of conversations children have with their parents play an important role in establishing memories, especially those that can be described verbally.

Did You Know ... A maître d' from a restaurant in Washington, D.C., claimed that the second time you visited his restaurant he could recite exactly what you ate the first time. He was also certain he could remember events from the first year of his life. But there was no way of verifying his childhood memories.

Of course some events in a child's life much earlier than this—even in the first few months— can still influence emotional life into adulthood, especially if those events are extremely stressful or traumatic. But these "memories" are not actively remembered. The youthful developing brain is likely to be more forgetful, too. It's busy rapidly assembling the structures and networks necessary for laying down permanent memories. And the creation of a memory involves multiple tricky steps: recording the memory, stabilizing it, and preparing it for storage in long-term memory banks. Making it through each of these steps is a risky process: if a memory isn't firmly implanted, it will be lost along the way, forever forgotten.

 **Did You Know . . .** Ultimately, forgetting some of what we experience is necessary: those rare humans who have the (dis)ability of remembering virtually everything are often, if not always, miserable. A woman known only as A.J. (for her privacy) has an extraordinary memory for dates and is not happy about it: "Most have called it a gift but I call it a burden. I run my entire life through my head every day and it drives me crazy!!"

So when you think back to the earliest events you can remember, the sketchiness of those few memories you've hung on to is likely due to the fact that your two-year-old brain was busy frantically building an effective memory system. Once you turned five or six, that system was pretty much in place, and your detailed reminiscences—true memories—began to be recorded from that time on.

Lest we forget!

Can we ever walk in a perfectly straight line?

ACCORDING TO LEGEND, people who are lost will walk in circles even though they're convinced they're making a beeline in the right direction. Nearly a century's worth of scientific experiments has established that when people have no landmarks or other reference points to use as guidelines, they will deviate systematically from a straight line.

Am I walking in circles?

Yes.

Yes.

Yes.

Yes.

Yes.

In 1928, Asa A. Schaeffer released a report called "Spiral Movement in Man." Despite what the title suggests, Schaeffer tested both men and women in a series of experiments by blindfolding them and asking them to move in a straight line, either by walking, swimming, rowing or driving. (The blindfolded drivers, one of whom was a nine-year-old boy, had copilots for safety).

In these experiments, most of which (with the exception of the swimming and rowing) were conducted in open Kansas fields, Schaeffer found that no one was able to navigate in a perfectly straight line. More than that, they often moved in such extremely curved paths that they performed a series of loops—that is, closed circles, the "spirals" in the title of the paper. Most individuals had a tendency to turn one way more often than the other, but, when given a direction to turn left or right, they could do so accurately. Ironically, when they were asked to trace the outline of a circle, some said they were sure they were mistakenly taking a straight-line path away from it, when they were actually walking an even tighter curve than had been requested.

Did You Know . . . According to Schaeffer's report, most of the blindfolded subjects found that following the path of a circle 131 feet (40 meters) in diameter felt exactly the same as trying to walk in a straight line.

Dr. Schaeffer either sketched or photographed the sometimes bizarre paths the subjects took, and was so convinced by the consistency of what he had seen that he proclaimed that these spiral paths were the same as those traced by microscopic single-celled animals like amoebas or paramecia. Moving in circles, he claimed, must be a feature of all living things. Schaeffer had done everything possible to remove any visual cues to direction (like having his subjects carry umbrellas to shield the sun) and acknowledged that the circles traced by people who were really lost in the woods would be much bigger, because they could not use distant geographic features as guides.

Despite the fact that Schaeffer appeared to be painstaking in his record keeping, there are always doubts that century-old science protocol meets today's standards. So, in 2009, Jan Souman at the Max Planck Institute for Biological Cybernetics in Tübingen, Germany, together with colleagues in Canada and France, extended the experiment by using GPS to track people— sometimes with a blindfold on, other times without—in a forest in Germany and in the Sahara Desert.

The subjects were allowed to walk freely for several hours. Those in the forest were able to maintain almost perfect straight lines as long as the sun was visible. But in cloudy conditions, they circled repeatedly, sometimes to the point that they doubled back and crossed their own paths. In the desert, those walking in the sun during the day slowly veered away from a straight line but never circled completely. One person who walked the desert at night was fine as long as the moon was visible, but started turning and eventually walked back toward the starting line when it disappeared behind clouds.

So the results are consistent: we do walk in circles, and they're smaller than you'd think. The bigger question is why.

One suggestion is that our legs are not exactly the same length—perhaps a few millimeters different—and that difference, however slight, makes us turn over long distances. But not only is this kind of a ridiculous idea, there is no evidence for it. Souman also tested the leg strength of his subjects and found no correlation between that and the direction turned. Sometimes the same person turned one way in one experiment and the opposite in the next.

Science Fiction! *The legendary dahu, a French mountain goat or antelope, was said to have legs significantly shorter on one side of the body than the other, allowing it to stand upright on steep slopes. Ascending a mountain was child's play for the dahu—until it reached the summit. Then it was stuck. Maybe that's why there are, and never have been, any dahus.*

Others have suggested that we move in spirals as a result of the Coriolis force, the same reason that hurricanes swirl counterclockwise in the northern hemisphere. But given that the Coriolis force isn't even strong enough to dictate the direction of water swirling out of the bathtub, there's no chance it could bend your path. And although you might expect right-handed or left-handed people to circle differently, both studies found this made no difference.

Back in 1930, Schaeffer argued that the spiral movements he observed must have their origin in the brain. That assumption remains common today. Some have suggested that the balance of body control between right and left hemispheres of the brain might be at the root of turning behavior.

Another possibility, supported by Souman and his colleagues, is that it simply isn't possible for us to walk in a straight line for any significant distance without cues. The sensory apparatus that keeps us on track—in particular, the eyes, the feedback from muscles as we move, and the balance system in the inner ear—gets overwhelmed by "noise," the sheer volume of incoming information, much of it irrelevant, making navigation difficult.

Without some sort of external guide, such as a compass, a grove of trees or the sun or moon, we're forced to rely on a vast amount of other incoming data, including the position and direction of our body, the passage of time, and the visual changes in the landscape as we move through it. If we can't update the system from moment to moment and focus on what's important, we deviate from a straight line. We just can't help it. That's how we get lost.

Why do we have five digits on each hand and foot?

IT'S NOT JUST US—it's the vast majority of mammals on earth: we all have five toes, or five fingers and toes. But why five? What's so special about that number?

There are exceptions: pigs only have four toes, rhinos and emus have three, deer and ostriches have two, and the horse one only. Each of those evolved to reduce the number of toes. That seems to be the trend: digits get subtracted, not added. But we can only guess what might have favored the loss of toes in each case. For instance, there's a theory that as the ancestors of horses moved from forests to the great plains, they acquired the need for speed, and shedding toes—all the way down to a single one—made them sleeker and faster. Or not: new research has shown that horse evolution isn't the simple straight line that's often portrayed, making bold statements about why digits were lost here or there inadvisable.

If you go back 460 million years, there weren't any land animals yet, but there were fish-like creatures that were able to breathe out of water. They were the first to explore the land. Their limbs were more fin than foot but showed signs of structural changes that would increase mobility in the water and also be helpful if moving on land—things like a more rigid shoulder joint.

These ancients, with names like *Acanthostega* and *Ichthyostega*, look like the rear end of a fish glued onto the front end of a salamander. Some had feet with eight digits, some seven or six. It was as if they were trying different numbers out, a grand experiment by nature to determine what to use over the next 400 million years. As the great migration from sea to land continued, and feet transformed to paws, five digits became the preferred number.

 **Did You Know . . .** Why did animals leave the water and crawl up on land? A new theory suggests they left so they could see farther. Water scatters light, and bigger eyes can't surmount that problem—but air doesn't nearly as much, so creatures emerging from water would have had wider vistas to explore.

You might be tempted to spread out your fingers, press them down on the table and think, "Yeah, five seems to be the perfect number for support," but we don't know exactly what ancient environments were like. If the ground was marshy, more digits would provide needed surface area. But if it was dry, speed was crucial, both for predators and prey, and fewer digits to lighten the load would be preferred.

Of course, no animal was anticipating anything; as time passed, natural selection was weeding out those that were disadvantaged in any way. So it wasn't long—at least in terms of geological time—before the six-, seven- and eight-toed species were gone, and five-toed ones were dominant.

Science Fiction! The panda's so-called extra "thumb" is actually a growth on one of the wrist bones, so it doesn't count as a digit.

Why is reducing the number of digits over time much more common than increasing them? Part of the reason is genetics.

In the early embryo at the time digits are being formed—in humans that's around four weeks—there's action everywhere. The sets of genes that control the development of digits are networked with other sets responsible for completely unrelated things, like the reproductive system.

The kind of dramatic genetic tinkering required to increase the number of digits would almost inevitably affect other systems. And so eight-, seven- and six-toed creatures mostly died out. (I say "mostly" because the Western clawed frog of Africa has, as the name suggests, a claw farther up the wrist from the other five digits, and scientists are arguing it's a digit in its own right.)

In most animals, genes operate smoothly to lay out five digits in the right locations on both feet and hands—Ernest Hemingway's six-toed cats are a rare genetic fluke. In the developing limb, a combination of genes either dampens or reinforces each other's activity in a precise way, with the result that patterns form in the tissues. In roughly the same way the zebra acquires its stripes, the limb first develops cartilage, then bone, in a five-digit pattern.

Did You Know . . . Are we losing a digit? Compared to the great apes, our baby toe has shrunk over the past 5 million years; sometimes it doesn't even have a toenail. Part of the reason is mechanical. We stride forward by landing on the heel of the foot, then transferring that weight forward, finally pushing off with the big toe. Our reliance on our big toe has lessened the need to maintain a power digit on the other side—the baby toe. And so it seems we're losing it, but how fast, nobody knows.

One last thought: our thumbs are clearly different from the four fingers, and it appears as if the genes active in forming the thumb are different from the genes responsible for the four fingers. We might owe this difference to those animals that first ventured out of the water onto land, and to the genes that helped convert fins to paws. But in the last 2 million years or so, evolution has further tampered with the thumb, converting it to the famous "opposable" thumb responsible for so much of our manual dexterity. We're not the only primates to have an opposable thumb, but we make the most use of it.

Did You Know . . . Our base 10 number system owes its existence to our five digits. But some cultures use base 20 by combining fingers and toes; others include the elbows, the nose, the neck and even belly button, and so have bases of 30 or more!

Why do my knuckles make that cracking noise?

WE'VE ALL SEEN—OR HEARD—IT. You watch someone interlace their fingers and push their palms away from them. As the person stretches their fingers, they're rewarded with a sharp crack before they settle back down to business. To many people it's a painful sound, but the sound is actually a sign of something being created, not destroyed.

The cracking noise that a knuckle makes has to do with bubbles in the joints. From the 1940s until now, most experiments designed to figure this out have looked pretty much the same: only the imaging equipment has changed. In most experiments, a volunteer has a finger wrapped and tied to a cord that can be pulled to apply tension to the finger. Any finger will do; in the 1940s, the middle finger was used for most experiments, but today it's the forefinger. A force is gradually applied to the finger as images are recorded of the joint between the last hand bone and the first finger bone.

At some critical point, as the finger is stretched, the force being applied crosses a threshold and there's a sudden, explosive noise. From the early 1970s to 2015, it was believed that the cracking was the sound of a bubble (really a void) in the joint imploding. As the bones stretch apart, the argument claims, the bubble bursts into existence, then, just as quickly, collapses.

But Greg Kawchuk and his lab at the University of Alberta have shown that it's the expansion of the bubble, not its collapse, that makes the sound. Kawchuk's images show that the birth of the bubble is just about immediate and that the suspected implosion of the bubble is more of a prolonged collapse. Their recordings of the crack show that at that exact moment of the sound, the space between the bones in the finger and the hand suddenly opens up, as much as doubling (0.04 to 0.08 inches, or 1 millimeter to 2 millimeters) in a tiny fraction of a second. That creates the bubble between the bones.

Before the finger is stretched, the finger bones are moving back and forth. They're in contact with each other, and any space between them is filled with fluid. When the bones are stretched, though, the separation is too sudden for more fluid to immediately flow into the space. So gases—especially water vapor and carbon dioxide—quickly enter the space from the surrounding areas instead, and a bubble forms instantaneously in the middle. High speed means big forces, and that's how the cracking sound is created.

Imagine trying to pull apart two sheets of wet glass stuck to each other; that sucking sound they make as they suddenly separate (which you can also make by squeezing the palms of your hands together and then suddenly separating them) is the same kind that's produced inside the joints of your fingers.

Any last doubts that this is a violent event were removed by a recent ultrasound study that revealed a bright flash of light, like fireworks, when a knuckle cracked.

 **Did You Know . . .** It can take anywhere from fifteen to twenty minutes for the finger joint to ease back to its original spacing so that it can be cracked again.

All of this cracking and separating sounds painful, so it's natural to wonder if cracking your knuckles can damage your hands. For a long time, especially when it was believed that the sound was created by the collapse or implosion of the void (bubble), it was thought that cracking your knuckles could damage your hands. After all, the same mechanism erodes ships' propellers: bubbles form around the edges and implode, causing a shock wave. Over time that causes the metal of the propeller to fatigue. But despite the number of theories out there, the most damage that's ever been attributed to hand cracking is from a group of knuckle crackers studied in the 1990s who suffered from reduced grip strength and some swelling of the hands.

Did You Know . . . When Dr. Donald Unger was a child, his mother warned him that cracking his knuckles would give him arthritis. Determined to prove her wrong, Unger proceeded to crack the knuckles on his left hand twice a day for the next fifty years. He left his right hand alone during that time as a control.

At the end of his trial, Unger reported in an article to the journal *Arthritis and Rheumatism* that there were no signs of arthritis in either of his hands. He argued that this disproved his mother's claim, and it prompted him to wonder whether he should continue to trust her admonition that eating spinach is good for you. Critics later argued that his sample size (one) was too small to draw any conclusions.

So, the sound of your knuckles cracking is the sound of the birth of a giant (relatively speaking) bubble in the joint. That probably doesn't make it any less painful for anyone listening.

History Mystery

*Is it true that right now we are breathing the same air
that Julius Caesar breathed?*

THAT LITTLE WORD "AIR" CAUSES CONFUSION. Air isn't a single thing; it's a mix of several gases. Nitrogen and oxygen make up 99 percent of the total, along with whiffs of argon, carbon dioxide, water vapor, and other trace gases. Every time Caesar exhaled, molecules floated off in all directions, dispersed by the wind. But could any of the atoms contained in one of Caesar's breaths from more than 2,060 years ago still be around for you to inhale? The answer is yes!

Beware the
~~carbon diox~~ides
of March.

Why is that? Atmospheric gases such as nitrogen, oxygen and carbon dioxide are made up of molecules, which are in turn formed from clusters of atoms. Over time, all molecules are subject to chemical and radiative forces that break them apart, but the atoms themselves aren't destroyed. Two single oxygen atoms that combine to make up an oxygen molecule might split up, then go on to attach to different oxygen atoms or perhaps latch onto a carbon atom to make carbon dioxide. It's an endless cycle. The exact molecules that Caesar breathed aren't likely to be floating around anymore, but the atoms that composed them are just waiting for you and your lungs.

 **Did You Know . . .** Julius Caesar was assassinated on March 15, or the Ides of March, in 44 BCE, a day that became notorious because of his murder. The air in his last breath out had about 4 percent less oxygen and 4 percent more carbon dioxide than it did when he inhaled it, because oxygen is used as fuel by our bodies, with carbon dioxide being the resulting waste. Caesar was stabbed twenty-three times by Brutus and dozens of other Roman senators, although apparently only one wound was fatal. More like a last gasp, then, not a breath.

Assuming Caesar was of average size, the amount of air he exhaled in a single breath would have been 30.5 to 61 cubic inches (about half a liter to 1 liter). A well-known formula tells us the number of molecules contained in that volume at an air temperature of 32 degrees Fahrenheit (zero degrees Celsius). Even accounting for Rome's slightly warmer, less dense air in March, a fair estimate is 1×10^{22}—or 10,000,000,000,000,000,000,000 molecules in just one exhalation.

Because more than two thousand years have passed since Caesar's death, we can assume that the air he breathed has been thoroughly distributed throughout the atmosphere. And, as you'd expect, there is an astounding number of molecules in the earth's atmosphere: 1.08 followed by 44 zeroes.

Okay, so Caesar exhaled 1×10^{22} molecules; let's say you inhale about the same, although if you take a really, really deep breath you might be able to take in twice as many—but with these numbers, twice as many makes almost no difference.

Those molecules were then diluted into a volume of air that was—and is—gazillions of times bigger. (A gazillion is not a real number, by the way.) On the other hand, all you need is one molecule of Caesar's to enter your lungs. Now it's simple division:

$$1.08 \times 10^{44} \text{ (molecules in the atmosphere) divided by } 1 \times 10^{22}$$
$$\text{(molecules in one breath)} = 1 \times 10^{22}$$

So you'd need to inhale (roughly) 1×10^{22} molecules of air to get one of Caesar's, and that is exactly what you're doing! So it's true: with every breath you take, there's a very good chance you're inhaling one of the molecules (or, to be pedantically correct, at least one of the atoms) that was part of Caesar's last breath. Keep breathing . . . big, deep breaths . . . and you are creating an even stronger bond with the great Roman leader.

Did You Know . . . Over the course of the average lifetime, a human being breathes around 500,000,000 to 700,000,000 times, taking about 900 breaths per hour. In comparison, the average dog breathes less than 200,000,000 times in its life—even though it's breathing twice as fast as a human, it lives a much shorter life.

Part 3
Animals

How do electric eels shock their prey?

THE ELECTRIC EEL IS AN AMAZING CREATURE on many different levels. Yes, it is basically a living battery, and it can use its electricity to detect, shock, and immobilize its prey or defend itself against predators. But it's even more bizarre than that: while it's a fish—big ones are up to six feet long—it has no scales and it breathes air. An electric eel must rise to the surface to breathe about every ten minutes or so.

Still, it's the battery that's the most interesting part. It's easy to ignore the fact that most living creatures, or at least those with more than one cell, run on electricity. Every single nerve impulse—and there are billions of them happening right now in your body—is a tiny jolt of electricity. The electric eel has taken this fundamental biological property and raised it to a high art.

The electric eel's body is not unlike a typical battery—say, the AA battery that powers so many household gadgets. Amazingly, about four-fifths of the fish's body is filled with stacks of tissue composed of specialized electricity-generating cells. Even though each cell produces only about one-tenth of a volt, there are thousands of them, and the layering allows the voltage to build from one end of the eel's body to the other.

Like a battery, there is a positive end, at the eel's head, and a negative end, at its tail. Just as you have to close the circuit in, say, your flashlight by flicking a switch to connect the two, the eel accomplishes the same thing by allowing its electrical current to flow from its the head to its tail through the water, which is a good conductor. Unlike a battery, though, when the electric eel releases current, it is sudden and powerful rather than steady and long-lived.

But not that powerful. Yes the electricity from electric eels is high-voltage, but the resulting current is relatively weak and short-lived. It gives a human a painful shock but one that's very unlikely to be fatal. In fact, the eel is a little like a Taser. Tasers deliver about nineteen high-voltage pulses per second. Electric eels produce many more, about four hundred per second. The effect is the same, whether on a human or fish: paralysis, since the electricity causes muscles to contract spasmodically. The Taser has a greater impact, though, because the shock is delivered through two metal darts that actually penetrate the skin and deliver it directly, while the eel has two disadvantages: it has to rely on its pulses traveling through the water, which offers resistance, and the pulse is emitted into the water without much targeting, so can't be anywhere near as focused—or concentrated—as the Taser.

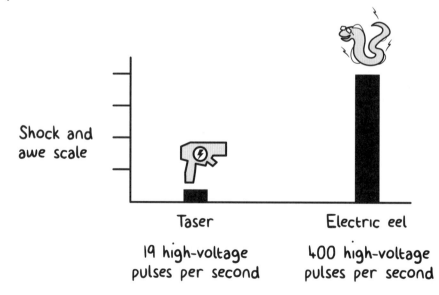

Shock and awe scale

Taser

19 high-voltage
pulses per second

Electric eel

400 high-voltage
pulses per second

That's not the end of the story. Ken Catania of Vanderbilt University has revealed that the electric eel hasn't just exploited electricity—it has refined it to an astonishing degree. It can turn the dial up or down, depending on exactly what it wants to do.

First, a low-power jolt allows the eel to locate its prey. When an eel emits one of these low-energy blasts, the muscles of a fish—even one that's hidden from view—involuntarily twitch. Eels are incredibly sensitive to any sort of disturbance in the water, so that twitch reveals the fish's presence.

Once the eel knows where a fish is, it can close in and power up the electricity and actually paralyze the fish; then it's game over. The eel's ability to temporarily paralyze a fish in this Taser-like way is obviously a pretty cool way to hunt and is extraordinarily effective: the blast only takes three milliseconds to take effect.

If this weren't enough, the electric eel also uses a clever technique to ramp up the delivery of electricity by curling its body in a C shape around any prey that is more difficult to subdue. The prey ends up trapped between the positive (head) and negative (tail) poles of the eel's body, with the electric current flowing directly between them both. This doubles the effective hit—already much higher than that delivered by a typical wall outlet—and most prey is unable to withstand that.

Did You Know . . . The great scientist and explorer Alexander von Humboldt traveled to the Amazon rain forest in 1800 to study electric eels. He was curious about how the eels hunted (electricity wasn't well understood at that time), and some local fishermen accommodated him by fishing for electric eels using horses.

According to Humboldt, the fishermen drove the horses into the shallow pools where the eels were hanging out. The eels' response was strange. Rather than fleeing, they shocked the horses, but not in the same way they would shock a fish. Instead, the eels reared up and actually made contact with the horses, delivering what looked like an extremely powerful shock, strong enough to fell some of the horses and drown them. After a while, though, the eels' electrical stores were exhausted and they could be collected by the fishermen.

One day, Ken Catania was using a metal net and protective rubber gloves to move his lab eels around. As he worked, he was amazed to see that the eels would sometimes leap out of the water, hit the net, and emit a series of incredibly high-voltage pulses. They probably feared the metal net was a predator: again, the eels use their electric discharges not only to kill prey but also to defend themselves against predators, and they likely determine the difference by the electrical conductivity of the unknown creature: small conductors are prey, large ones are predators.

When an electric eel sees a potential predator (like a large metal net), it can employ its standard approach and rely on water to conduct its electricity, but that might not provide enough current to deter an attack. By making physical contact with the attacker like a Taser, the eel's electricity would flow directly into the attacker's body, rather than through the water, making the shock substantially more powerful. And the higher up on the body the eel gets, the more electricity it appears to deliver, hence the leap out of the water, both in Catania's lab and Humboldt's pond. It would be a useful defense when creeks and streams become shallow in dry seasons, and prey or predators aren't fully submerged in the water.

Everything about the electric eel shows that the animal is not just an electric generator but a fine-tuned detection and killing machine. One thing that's still unclear, though, is why electric eels don't shock themselves. Maybe they're well grounded. At least, that's the current story.

I hope you found this electrifying.

How can a mongoose survive a cobra's bite?

In Rudyard Kipling's *Jungle Book*, the mongoose Rikki-tikki-tavi realizes, "If I don't break [the cobra's] back at the first jump . . . he can still fight." He looks at the thickness of the cobra's neck below the hood and knows that's too much for him and that a bite near the tail would only make the cobra savage. "It must be the head . . . the head above the hood. And, when I am once there, I must not let go."

Rikki-tikki's planning a surprise attack, but even so, it's risky. The venom of a king cobra, for example, can kill a person in half an hour. In fact, the amount in one bite from a king cobra can kill twenty people: it can deliver about a quarter of a shot glass of venom in one strike, more than any other snake. Happily for us, cobras prefer to avoid contact with people; their most common human victims are snake charmers.

And while mongooses are notorious tough guys—sturdily built, low to the ground, hard preda-tors—Rikki-tikki is confident for good reason: he has biochemistry on his side.

The secret lies in the venom and how it works. A king cobra's venom looks like it was assembled by a fussy yet nasty mixologist: it's a combination of many different types of venom that target different organs, all in one convenient dose. One of those venoms is common to many species of snakes: alpha neurotoxin. It exerts its deadly effects where nerves meet muscles, but at an ultra-microscopic level, where one molecule comes into contact with another.

Nerves signal muscles to contract. There's a tiny gap between them, and when an impulse sweeps along the nerve and reaches the end, the nerve releases millions of molecules called neu-rotransmitters. These drift across the gap—a mere millionth of an inch—and plug into special receptors sitting on the surface of the muscle. Then the muscle contracts.

Once that happens, the transmitter molecules have to be cleared away to allow the process to be repeated. There's another molecule whose job it is to do that.

If this sequence—nerve impulse, transmitter release, muscle contraction, removal of transmit-ters—weren't happening all over your body all the time, you couldn't breathe, much less move.

Oh, for goodness' snake!

Alpha neurotoxin targets this usually smooth-running system. Like the neurotransmitters, it's shaped to fit into the receptors on the muscle cell; unlike the transmitters, it doesn't get removed. That muscle cell will never contract again. As the venom spreads through the body, more and more contact points between nerves and muscles are blocked. It's said that even an elephant can be killed by the typical dose of king cobra venom.

So what does Rikki-tikki-tavi have going for him? Evolution. Over long periods of time, and with a great debt owed to many dead mongooses, changes to the mongoose's receptor molecules have made them resistant to venom. The venom molecules simply can't bind to the receptors the way they do with other animals, but the mongoose's neurotransmitters still can. So while the mongoose's speed and thick fur are both protective, once bitten, it has another powerful mechanism for self-defense. And you don't have to look far to see other, similar examples of creatures that have developed immunity. The honey badger has a good reason for its attitude. ("Honey badger don't give a s--t," it's been said.) The honey badger also has altered receptors, and that makes it a snake eater rather than snake prey.

Did You Know . . . Ironically, the king cobra is resistant to snake venom. Why? Because snakes are the cobra's primary prey and it has to be protected against them.

The final weapon in this array of molecular blows and counterblows is antivenom. To make antivenom, small amounts of snake venom are injected into domestic animals, like horses or sheep, and the animals' immune systems react by manufacturing antibodies. Those antibodies are then collected from their blood and concentrated. Once injected into a snakebite victim, the antivenom and venom molecules meet in the bloodstream and enter into what becomes a fatal embrace for the venom. It can't escape the antibody, making it impossible for the venom to attach to the muscle receptors, which means life goes on.

The mongoose's story is a classic example of evolutionary moves and countermoves. Right now the mongoose holds the upper hand in the evolutionary battle, but the fact that the cobra's venom contains so many different molecules with wildly different micro-architectures leaves the door open for one of those molecules to mutate into a deadlier form. When that happens, the tide will begin to turn.

What's the difference between falling toast and falling cats?

FOLK WISDOM TELLS US THAT TOAST ALWAYS FALLS butter-side down and cats always land on their feet. That sounds simple, but the science behind those claims is a little more complicated.

Let's take toast first. There is solid scientific evidence that if your buttered toast slips off the table, it will almost always land butter-side down. The word "slips" is key here. We're not talking about a violent arm gesture that sweeps the toast off the table and across the room. The toast's last moments on the table are crucial. Imagine watching in slow motion as the toast extends beyond the edge of the table. At first the table holds the bread up; however, when the center of gravity of the slice (roughly in the middle) extends out over the edge, the majority of the toast's mass is unsupported and it will start to tip over, only momentarily restrained by adhesive forces that help it cling to the tabletop. Tipping causes the toast to start to rotate as it plummets toward the floor.

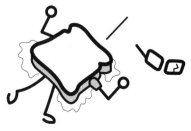

It's going to be one of those days.

To end up butter-side up, the slice of toast must turn either less than 90 degrees or more than 270 degrees. (I'm ignoring the near-impossible result that the toast lands on its crust and stays vertical!) Anything in between—a 180-degree turn or even anything close to that—will ensure butter-side down.

The butter itself doesn't play a significant role in this equation, because its mass is slight compared to the mass of the toast. And while melted butter does have the potential to change the aerodynamics of a piece of toast by making one side slicker than the other, those changes are so small that they don't have a significant influence.

Given that, what determines how much the toast will rotate on its way down? The height of the table! Most tables are about 2½ feet (a little over 75 centimeters) high—a comfortable height for a typical sitting human. That makes it extremely likely that the fallen toast will rotate 180 degrees (or at least between 90 and 270) and end up butter-side down.

There have been two studies done on this issue of turning toast. In one, an astounding 21,000 pieces of toast were dropped. In the other, the percentage of butter-side-down slices of toast ranged from 62 percent to 81 percent—impressive numbers.

 TRY THIS AT HOME! If you want to maximize the chances that your toast lands butter-side up, you have two options. One, change the height of your table: Build a higher table, preferably roughly nine feet (about three meters) high, or—as that's not a practical approach—sit at a much lower table, such as those found in Japanese restaurants. Two, eat much smaller pieces of toast. The physics argues that if each slice were about an inch across, you'd stand a much better chance of your little toast square ending butter-side up.

And what about cats? Cats are able to land on their feet within a few weeks of birth, and they rarely land upside-down, even if they're dropped that way from as little as 12 inches above the floor. (Please don't try this at home!) But while the physics of butter-side-down toast is pretty straightforward, the mechanics of a cat landing feet first is much more sophisticated.

The challenge the upside-down cat faces is the same as that of a gymnast or diver in midair. Physics imposes limits on how far you can twist or turn when you don't have anything to grip on to or push off from. This doesn't mean that it's impossible to move in midair: we've all seen figure skaters accelerate a spin by bringing their arms as close as possible to their bodies. But there are laws of physics, and a falling cat is restricted to working within those unbreakable laws.

Did You Know . . . Cats use a combination of their eyes and inner ears to analyze their position in the air. A blind cat can still land on all fours; so can a cat missing the inner-ear organ of balance. But a cat missing both cannot.

A cat does have advantages over a human: it has more vertebrae in its spine than we do (thirty versus twenty-four) and it lacks a collarbone. Both enhance its ability to twist, and twisting is the key to landing on its feet.

The first thing the falling cat does is arch its back like the stereotypical Halloween black cat. By doing so, the cat actually divides its body into a front half and a back half (in terms of physics, at least). Then it twists its front half and tucks its front legs in—like a twirling figure skater—so that it's able to twist further. At the same time, the back end twists in the opposite direction and, crucially, the cat extends its hind legs so that its back half doesn't rotate as much.

This brings the front part of the cat's body right-side up. It then reverses these actions, extending its front legs and twisting them a little in the opposite direction while rotating its back half more quickly by tucking its hind legs in. Within a fraction of a second, the cat is entirely right-side up.

Science Fiction! Cats do not need their tails to land on their feet; even Manx cats, which are tailless, can do that.

But there's more. Cats seem to be exceptions to one rule of physics: that of air resistance. The biologist J. B. S. Haldane once described the rules of air resistance in terms of animals: "You can drop a mouse down a thousand-yard mine shaft; and, on arriving at the bottom, it gets a slight shock and walks away, provided that the ground is fairly soft. A rat is killed, a man is broken, a horse splashes."

But what about a cat? A 1987 study by veterinarians in New York City claimed that cats that fall from great heights have incredibly high rates of survival, much greater than you'd expect.

That might be partly due to the fact that cats' terminal velocity while falling is only half that of humans, and that's thanks to their ability to twist while falling. Once cats have righted themselves in midair, they extend their legs sideways to maximize their air resistance and glide to the ground the way flying squirrels do.

In the age-old rivalry between canines and felines, score one for cats here, because dogs can't right themselves in the air the way cats do. That makes sense from an evolutionary perspective: dogs evolved from ground-dwelling ancestors, whereas cats were, and still are, tree climbers. Toast still has a few eons to go before it catches up.

Score one for the felines!

How do octopuses camouflage themselves?

You can have chameleons—when it comes to camouflage, I'll take octopuses, squids and cuttlefish any day. Collectively known as cephalopods, all three can disguise themselves by changing their appearance and fading into their background.

Imagine how tricky that is: a predator is standing across from you, and your only escape is to replicate your surroundings so the predator's eyes slide past. I'll admit there is some pretty amazing advanced camouflage gear for hunters and soldiers, but if you're wearing it, you'd better hope you're in front of camo wallpaper so you blend right in. Cephalopods, however, move past multiple backgrounds from moment to moment, and predators approach them from all directions, so they have to be able to adopt their pattern and color in a split second or they're lunch.

So how do they escape detection? Because they have brilliant biological tools at their disposal. Their skin amounts to a high-definition, pixelated, multicolored electrical sheet, with three layers of specialized cells. One layer is made up of colored cells called chromatophores, which are controlled directly by the creature's brain. Each chromatophore has muscles that contract and pull, expanding its surface area by as much as five hundred times, dramatically increasing the amount of color the skin displays. Considering there are millions and millions of chromatophores in a cephalopod's skin—at least 10 million in a cuttlefish—this is a potent system.

Below the chromatophores are two layers of reflector cells, both of which are colorless. The first layer reflects light in the same way a soap bubble does. You might have noticed that as a soap bubble drifts through the air, it changes color, from red through orange to blue and eventually, just before it pops, black. Those changes are the result of the soap film getting thinner and thinner and changing the wavelength of the light it reflects. Cephalopod skin has cells with stacks of plates in them; the thickness of the plates, together with the width of the spaces between them, creates soap bubble–like interference with light, which generates an array of colors. The reflector cells in the second layer are brilliantly reflective spheres, like tiny disco balls, that reflect light of all wavelengths from all directions.

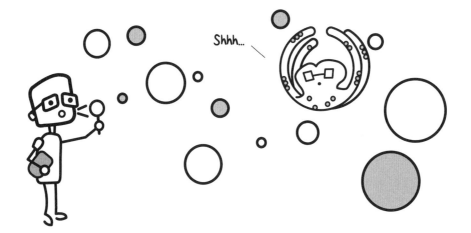

By adjusting the size of its chromatophores, a cephalopod can produce three patterns: uniform, mottled and disruptive. The uniform pattern is very fine grained and is produced by shrinking the chromatophores. An animal might use this as camouflage against sand. The mottled pattern is coarser, like a gravelly sea bed. The disruptive pattern is a dramatic display with big chunks of different colors: Think of a seabed dotted with corals and rocks.

So how do cephalopods choose which of the three display variations to use? Experiments with cuttlefish show that sometimes they match the background on which they're resting. Other times they "masquerade" as a rock or other feature somewhere nearby.

Sometimes by doing that their outline seems to change and they don't look like a cuttlefish anymore. The challenge for any of these camouflage-dependent animals is to take into account the kind of predator it will most likely encounter. This has been called the "point of view" predicament: the predator sees a different scene from the prey. For instance, the octopus has to use the visual information it gathers from where it is to create camouflage that will work for both a fish attacking from above and for a moray eel attacking from the side.

A different kind of complication arises for free swimmers, like squids and cuttlefish. If a predator approaches from below, the cephalopods are backlit by sunlight; but even though their bodies are partially transparent, their internal organs aren't. In this situation, the reflector cells come into play, channeling light through their bodies, making their insides lighter and harder to see.

Science _Fact!_ All cephalopods are intelligent (although it seems octopuses might be the smartest of the lot).

And what about that color-blindness? The colors produced by chromatophores are mainly yellow, orange and dark brown (and their combinations), which are a good match for the environments these animals have colonized. So it's not so much that cephalopods see those colors and adjust their chromatophores to match but rather that they sense brightness and shading and depend on chromatophores that are already tuned to the approximate colors.

There's also some evidence that texture (fine grained or coarse) rather than color provides the best protection. In fact, some cuttlefish are able to perform camo magic in near darkness.

Cephalopods also use camouflage as a means of communication. What could be more effective for sending messages than sets of flashing colors? Besides, those reflector cells also polarize light, and most fish—among them these animals' predators—can't perceive polarized light, but cephalopods can. Not just a visible communication but a covert one as well!

Did You Know . . . Squids have such exquisite control over their displays that they can have a polka-dot body, a striped fin and dark tentacles all at the same time.

Predictably, the military is interested in the mechanics of cephalopod camouflage and hopes to one day apply them to soldiers' uniforms. And understanding the cephalopods' ability to change color and pattern could even lead to new cosmetics and product lines one day, too.

Can an elephant jump?

THERE'S A MYTH THAT ELEPHANTS are the only animals with four knees. If that was the case, elephants might be great jumpers. But they don't have four knees: the bones in their front legs and our arms, for example, match one to one. They're the same kind of bones. So the question of whether elephants can jump has nothing to do with the number of knees they have; it's a question of force and pressure.

Huge animals like elephants, hippos and rhinos have legs that are thick like columns, much thicker relative to their bodies than, say, the legs of a cat. The bigger an animal is, the more its proportions change. An elephant is about thirteen times taller than a cat but weighs eight hundred times more. If it were the same proportions as a cat, that massive weight would snap its skinny little legs in two.

I think I can!
I think I can!

So how does the elephant support its enormous weight? With thicker legs. A cat's thigh bone is around 0.4 inches (about 1 centimeter) in diameter, but the elephant's is ten times wider (4 inches, or 10.2 centimeters). A larger diameter translates mathematically into a much bigger surface area, meaning that the weight-bearing surface of the elephant's leg bones is about twenty-five times that of the cat's. It's enough that the elephant doesn't break its legs, but the weight still puts a lot of stress on those elephant bones and makes the elephant, as large as it is, more fragile than a cat. That might be why, no matter how fast it moves, an elephant always keeps at least one leg on the ground. The impact from landing after being in midair might be too much.

But there are peculiarities about the elephant's legs that suggest we shouldn't rule out the possibility of jumping. For instance, most four-legged animals switch through a set of different gaits as they accelerate. Typically, the animal starts by walking, then accelerates to trotting, pacing and then galloping. By the time most of those animals are galloping, they spend a significant amount of time with all four feet in the air.

 **Did You Know . . .** Human sprinters spend most of a 100-meter race in the air, with both of their feet just above the track.

But elephants don't seem to switch gears like that. Instead, they just keep walking faster and faster. Elephants can get up to about 15 miles per hour, but even at that top speed, they're keeping at least one foot on the ground. At high speeds, the elephant's front legs are essentially still walking: it's called the "vault" pattern, because the leg is planted and the animal moves over it in the same way that a pole-vaulter plants his or her pole and flies over it. But its hind legs are bouncing a little on impact and behaving more elastically than you'd expect from an animal of this size. Other massive animals, such as rhinos, gallop like regular four-legged animals, but even rhinos are actually quite a bit smaller than elephants, so it might be that elephants are just too big to gallop.

There's another difference between the fore- and hind limbs: typically, quadrupeds use their back legs to push themselves forward, while their front legs are the brakes. But Dr. John Hutchinson of the Royal Veterinary College in the UK has been studying elephant locomotion for decades, and he's shown that elephants can use any leg for both braking and propulsion. Elephants are nature's 4x4s.

It's not just the force of body weight on the bones that affects an elephant's ability to jump. The other parts of the leg play an important role, too. Good jumpers have flexible ankles, strong calves and sturdy Achilles tendons, and there's little to show that elephants have jump-worthy versions of any one of those. Most animals that can jump well do so to evade predators, but elephants don't really have any. Elephants' musculature has evolved to propel them forward, not upward. But even if an elephant isn't built for jumping, could it leap if it was forced to?

No one's ever seen it, and they might not have the muscular power in their lower legs to propel themselves off the ground. And even if they did launch themselves, the impact forces of takeoff and landing would be huge, all of which gives good reason to doubt that elephants could—or even would—jump.

They call me a flatworm, but
I'm smarter than that.
I'm a bookworm.   ———————

Can worms digest each other's memories?

Perceptive Platyhelminthes

BELIEVE IT OR NOT, THIS IS A SERIOUS SCIENCE QUESTION, full of controversy. The story goes back to the 1960s and experiments conducted with flatworms known as planarians. They're simple worms: a centimeter and a half long, as flat as the name suggests, with a tiny brain that is surprisingly capable of learning. But it's this flatworm's odd history as an experimental animal that makes it so special.

If you cut a planarian into two halves, each half will reform itself into a whole new animal. Knowing this, a somewhat eccentric scientist named James V. McConnell claimed that if a planarian were trained to avoid a light, then was cut in half, both the head end and the tail end, after regenerating into two new animals, would remember the training. That the head (with its brain) would retain learning made sense, but the tail? That was unbelievable, and it went against every theory of memory that existed.

Did You Know . . . Planarians already cut into pieces have been in space in the International Space Station? When they returned, one of them had regenerated two heads, one at each end of its body. When, back on Earth, both those heads were cut off, the body generated another two. No one has ever seen this happen on Earth, and why it happened is unclear.

Scientists were disturbed by McConnell's claim, because memories were thought to be preserved as special electrical circuits among brain cells. When something dramatic happens, the brain reacts with pulses of electricity, and if the event is important enough, a memory of it is preserved by a set of neurons arranging themselves to keep that circuit alive. McConnell's suggestion that memories were instead some kind of material or molecule that could be found throughout the body seemed, well, crazy.

It didn't help that McConnell was a little different from most scientists. Not only did he publish his experiments in his own cartoony journal called the *Worm Runner's Digest*, he even claimed that he'd gone one step further and had been able to transfer memories between planaria by encouraging untrained worms to eat pieces of worms that had been trained to avoid the light. That's right: learning by cannibalism.

By the mid-1960s, McConnell's research came to an end, doomed both by the inability of his colleagues to come up with evidence to support his claims and by a general intolerance to his out-there ideas. McConnell's flatworm experiments fell off the radar, and he died in 1990.

But a few years ago, a surprising report on flatworm experiments by scientists at Tufts University gave dramatic new life to McConnell's ideas. Michael Levin and Tal Shomrat trained planaria to become accustomed to their surroundings (specifically the texture of the surface on which they were fed), then cut the worms' heads off, waited for them to become whole again and placed them back in the feeding area. The flatworms that had been trained to recognize the unique feeding surface, even though they now had new heads, were quicker to begin eating than those that hadn't.

They were quicker, but they did need a brief reintroduction to the environment in which they were trained before they showed significant advantages over their untrained colleagues. It's not

exactly clear why that should be, but Levin and Shomrat suggest that memories, or at least a molecular reflection of them, are somehow stored in nerve tissues other than the brain—such as the part of the planarian nervous system that extends from the brain to the end of the tail—or even in other kinds of tissues. So, as a new head grows, memories are incorporated inside the new brain or are impressed upon it.

One way that might happen is through epigenesis, or the manipulation of genes by other molecules, especially proteins. If that was the case, Levin and Shomrat wondered if this ability might have implications for the restoration of memory in humans by using stem cells containing a person's memories. But, as amazing as that sounds, we're still a long way from understanding the true nature of memories.

I don't think these experiments are going to persuade scientists to change their minds about the nature of memory overnight. But the idea that, at least in some animals, memories might be stored in places other than the brain refuses to go away.

Could humans ever hibernate?

WHEN WE THINK OF HIBERNATING, we think of bears fattening up, finding a den and sleeping all winter. But there are many ways to hibernate, and there may be some compelling reasons why humans would want to do it. The most obvious application of hibernation is for keeping patients in a slowed-down state for transplants, for prolonged surgery and recovery or even for space travel.

I can't *bear* the thought of winter.

A typical bear hibernates for anywhere from five to seven months. During that time, its core body temperature drops 8 degrees Fahrenheit (5 degrees Celsius) and stays there for weeks at a time. The body shuts down to about two-thirds of its normal activity and the heart slows from forty beats per minute to ten to fifteen. Even though they're hibernating, bears still burn a few thousand calories a day.

They can lose up to a quarter of their body weight during hibernation, almost all fat, and yet they suffer no bone loss or muscle wasting (whereas bedridden humans certainly do). Bears don't pee or poo when hibernating. Their kidneys nearly stop but don't fail, and they don't accumulate deadly levels of chemicals.

Did You Know . . . When bears emerge from their hibernation in the spring, they don't eat or drink much for the first couple of weeks, even though they've taken in nothing for months. That changes quickly, though. By midsummer a good-sized bear consumes 5,000 to 8,000 calories a day, and grizzlies, just before hibernation, take in 1,000 calories an hour, twenty hours a day.

Bears grab our attention, but when it comes to hibernation they're not nearly as impressive as some smaller animals. Arctic ground squirrels in Alaska hibernate for anywhere from eight to ten months. As they hibernate, the temperatures in some parts of their bodies can drop as low as a degree or two below freezing. That's impressive, but some animals can survive being totally frozen. For example, if wood frogs are dug up in the winter, they are exactly like ice cubes: if you dropped one on the floor, it would shatter. Ground squirrels don't turn into chunks of ice like this, though. Instead they're just supercooled: the water in their bodies stays liquid instead of freezing. This is a precarious state: you can put a container of water in the freezer and sometimes it will remain liquid until you jostle it or even just tap the side with your finger. Any disturbance like that triggers the instantaneous formation of ice. Rats and hamsters can be supercooled, but if they're left in that state for more than an hour, the water in their bodies will start to turn to ice and crystallize and they'll die. Somehow, Arctic ground squirrels can maintain that awkward state for days. It's not clear yet how they do this; they might be able to clear their bodies of any particles around which ice crystals could form, but no one knows for sure.

These squirrels also differ from bears in that they wake up every few weeks to raise their body temperature, move around a bit, pee and poo, then return to hibernation. These brief awakenings must be necessary for the animal's survival, because it costs a hibernating animal a lot of energy to return to normal life and then slow down again. Maybe these awakenings have something to do with the extremely low temperatures they reach.

Looking at bears and squirrels, you might assume that hibernation is all about escaping the cold. But that's not always true. The fat-tailed lemur—an animal living in Madagascar, one that is more closely related to us than either bears or squirrels—hibernates for about seven months of the year even though the temperature in its tree den can range from 50 to 86 degrees Fahrenheit (10 to 30 degrees Celsius) during that time. The seven months that the lemur is holed up are the driest months, when food is scarce, and apparently it's just not worth it for the lemur to patrol the forests searching for it. Instead, it prepares for hibernation by practically doubling its weight in the plentiful months.

The fact that animals' bodies remain healthy through the dramatic changes of hibernation is amazing enough, but if we want to know if humans could hibernate, we have to focus on the brain: it's the most energy-hungry organ, and a hibernating animal does everything it can to reduce energy consumption.

Science Fiction! *Hibernation is not sleep. In fact, hibernation might actually deprive animals of the kind of sleep they need most. When Arctic ground squirrels break out of hibernation every few weeks, they use a big chunk of their non-hibernating time to sleep. It seems that sleep is something a hibernating animal can't afford to do, because sleeping burns too much energy.*

A slowing metabolism has dramatic effects on brain cells. As a hibernating animal's body slows, the branching that allows each brain cell to communicate with thousands of others shrinks and retreats, and the skeletal system that supports them starts to collapse. Some of the changes seen in hibernating brains actually mimic destructive changes seen in the brains of Alzheimer's patients. But when the animal wakes, its brain kicks off an amazing bout of activity, reestablishing connections and restoring skeletal frameworks. It's not completely clear whether these changes affect the animal's memory or not.

Did You Know . . . A Japanese hiker named Mitsutaka Uchikoshi survived twenty-four days on the side of a mountain without food and water after he passed out, having suffered a broken pelvis as he was hiking. When he was found, his body temperature was 71.6 degrees Fahrenheit (22 degrees Celsius). He should have been dead, but he survived virtually undamaged. Although this wasn't the same mechanism that a bear or an Arctic ground squirrel goes through, doctors were quick to say that his survival was due to something like "hibernation."

We have not evolved to hibernate, so some features of hibernating animals, like maintaining a heartbeat at 33.8 degrees Fahrenheit (1 degree Celsius) are impossible for us. The human heart fails when body temperature drops below 68 degrees Fahrenheit (20 degrees Celsius). Like the lemur, we'd have to arrange hibernation at a reasonable temperature. And we would have to figure out ways of ensuring that we would wake up from hibernation every few weeks—as the Arctic ground squirrel does—so that we could eat, drink, pee or poo (or all of these). That might also help with the issue of sleep deprivation. If our brain cells started to lose connections and our skeletons started to fall apart, we'd need to be absolutely sure that everything would recover when we woke.

Even so, human hibernation might not be impossible. There could be a way of shortcutting millions of years of evolution to provide a safe, human-only form of hibernation. But the subtlety and sophistication of the mechanics of it suggests it's a long, long way off.

History Mystery

Was Atlantis a real city?

WE CAN ALL CONJURE UP FABULOUS IMAGES OF ATLANTIS, complete with gold and silver buildings—all underwater. But was there ever such a place? And, if so, where on earth was it?

If the question is understood as: "Was there an island of untold riches that was swept underwater by some cataclysmic event, never to be seen again?" the answer is "Not a chance." But if the question is "Does the legend of Atlantis refer to a real disaster that brought down a kingdom?" then the answer might possibly be "Yes!"

Hundreds of authors have written about Atlantis, but we can trace the legend back to one man: the ancient Greek philosopher Plato. One of the founders of Western philosophy, Plato lived during a glorious period in history. His mentors included both Pythagoras and Socrates, and Aristotle was one of his students. All of these great thinkers would easily be inducted into the Philosopher/Scientist Hall of Fame, if there were such a thing.

Philosophy begins in wonder
(and a hall of fame).

Around 360 BCE, Plato wrote two essays in which he described conversations. In the essay called *Critias* (named after an Athenian politician), Atlantis is described in detail. Critias explains that he'd heard about Atlantis from various people and that these storytellers left no doubt in his mind as to the island's splendor. It was said to have existed where the Mediterranean Sea meets the Atlantic Ocean, just beyond the Pillars of Hercules. The northern pillar is thought to be the Rock of Gibraltar; there's controversy about what the corresponding pillar or peak on the African side might be. Still other accounts suggest the pillars might actually be stars, and if that's the case, finding Atlantis just got a whole lot more difficult.

Atlantis is supposed to have existed nine thousand years before Plato wrote about it. Plato describes it as having a huge city, with 6.2-mile (10-kilometer) canals that allowed ships to enter its inner harbor. The city was laid out in concentric circles, with canals between and connecting each circle. There were bridges, tunnels and a palace with an ivory roof, silver-coated walls and gold statues, not to mention gymnasiums, a racetrack and dockyards. The island was rich in vegetation, with freshwater springs and minerals worth more than gold. Plato's description was so over-the-top as to be, well, unbelievable.

Real or not, the kingdom didn't last. According to Plato, "there were earthquakes and floods of extraordinary violence, and in a single dreadful day and night . . . the island of Atlantis . . . was swallowed up by the sea and vanished."

Somehow it's not surprising that this ancient description of Atlantis created a two-millennium-long obsession with discovering the sunken island's location. Is it at the mouth of the Mediterranean? There's no evidence of anything there. Could it be somewhere else, such as in the Canary Islands, in the Sahara Desert, in the North Sea, on the island of Bimini or in Antarctica? Each of these places has been suggested, but no trace of the island and its ancient city has ever been found at these sites. Also, nine thousand years before Plato, there were no such magnificent cities anywhere on earth: clay-and-straw-brick houses at Jericho were about the best you could find.

But if Plato was referring to a real disaster that struck down a civilization, then there might be evidence of that. Around 1650 BCE—1,300 years before Plato wrote about Atlantis—a giant volcanic eruption in the eastern Mediterranean leveled the island of Thera (now called Santorini). The blast was so powerful that almost half of Thera collapsed into the sea, leaving behind an irregular ring of much smaller islands. A mix of lava and gas hit the water at speeds of hundreds of kilometers an hour, creating tsunamis with estimated wave heights of 33 to 49 feet (10 to 15 meters) and a towering column of ash that spread over the Mediterranean, leaving deposits several centimeters thick even on the mainland, hundreds of kilometers away. Rafts of frothy lava (pumice) covered the surrounding sea.

The eruption is almost as sensational a story as the original account of Atlantis—better in some ways, because we know for a fact that it happened. However, even though the town of Akrotíri was completely buried under volcanic matter, no bodies have been found there, suggesting that the citizens had early warning and were able to flee well in advance. And Akrotíri was not anything like the magnificent city that Atlantis was rumored to be.

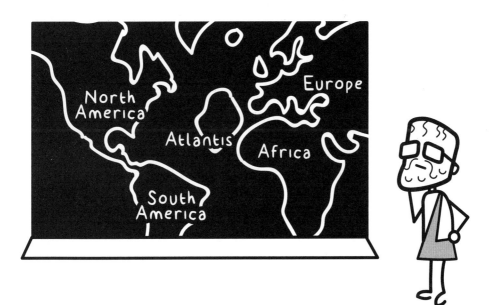

But if not Thera, what about the island of Crete? It lies a little over a hundred kilometers, or about seventy miles, southwest of Santorini. At the time of the eruption, it was home to the Minoan civilization. The myth of the minotaur (a beast with the body of a man and the head of a bull) and the labyrinth (his home) have come down to us from the Minoans. After the eruption, the Minoan civilization collapsed. Many archaeologists have argued that it's the collapse of Minoan civilization that Plato was referring to when he wrote of Atlantis. There's significant doubt, though, about whether the eruption decimated the Minoans. For one thing, the ash seems to have fallen only on the eastern end of the island, and not much of it at that.

But what about that tsunami: Did it hit the Cretan coastline hard? Evidence, such as pieces of the same pot found in different rooms of a shattered house, would suggest the tsunami was chaotic and powerful, but recent studies show that some places along the exposed coast were minimally affected. And the cause and effect are murky because Minoan civilization took hundreds of years to wither away. I guess it's still possible to argue that the Minoans were severely disrupted by the combination of the damage the tsunami had on shipping and the damage the ash deposits had on agriculture, and that those factors contributed to the slow demise of the culture. But that's a pretty watered-down version of the story, and it still doesn't account for the lavish kingdom Plato described.

So what was Plato writing about? If not a lost island kingdom in the Atlantic Ocean, then a volcanic eruption closer to home in the eastern Mediterranean? And did he base his Atlantis tale on the entombment of a single unremarkable village, Akrotíri, on the island of Thera? If that's the case, the lavish descriptions of Atlantis are a tribute to his imagination but not to reality.

Part 4
Weird Science
& Machines

How do stones skip?

WHEN IT COMES TO SKIPPING STONES, there are three kinds of people: those who love it as a cottage activity, those who take it seriously enough to try for world records, and those who try to understand the physics of it. It's incredible that there are still physicists focused on this question, because people have been skipping stones for ages.

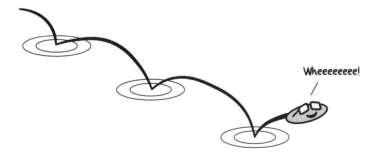

Wheeeeeeee!

The ancient Greeks referred to skipping stones across water millennia ago. But the first well-attested records are from the early seventeenth century, when King James I of England, who ruled from 1603 to 1625, amused himself by skipping gold sovereigns across the river Thames (he was king, after all). Gold sovereigns weigh about 8 grams, or a little more than a quarter of an ounce. That's extremely light, and as anyone who's skipped a stone before knows, when a rock is too light, it doesn't stay horizontal long enough to skip well. I bet if King James was somewhere in central London, he was throwing at low tide, making it easier to get the coins across the river.

The first scientific account of stone skipping comes from the lab of Lazzaro Spallanzani in the 1700s. Spallanzani didn't just watch people skip stones and guess what was happening, though. Instead, he dropped and threw stones and recorded the results. He even managed to get stones airborne by grasping them between thumb and forefinger and sweeping them along the water. Some of his observations were pretty straightforward: for instance, he showed that you need a flat stone and that it has to hit the water on its flat side, not its edge, if it's to skip. He then added the crucial observation that a stone will skip much higher if it hits the water at a slight upward angle.

 **Did You Know . . .** Spallanzani was an accomplished guy. Not only was he the first to show that bats echolocate, he also did crucial experiments to disprove the concept of spontaneous generation—that life could simply erupt from non-living matter. He showed that broth left exposed to the air quickly spoiled and became infested with bacteria, but that the same broth sealed off from the air did not. No surprise to us, but a revelation at the time.

Most of those points are obvious to anyone who's skipped a rock before. But Spallanzani dove into the details. He fired a lead shot at a low angle across the water. From what he witnessed, he argued that when a stone thrown (or a bullet fired) at a low angle strikes the water, it creates a depression, rides down the near side of the divot and then slides up the other side and into the air. It's an astonishing observation, because it would have required a ton of patience and a keen eye.

Spallanzani's findings weren't referred to again until the twentieth century, when modern technology allowed for more precise measurements. In a high-speed video shot in the late 1960s, researchers were finally able to view in detail what Spallanzani had observed with his naked eye. The video showed that a skipped stone does indeed hit the surface and push down the water, creating a wave in front of it. The rock then continues to skim forward along that wave, the front edge slanting higher and higher until it reaches an angle of something like 75 degrees (for comparison, the Tower of Pisa stands at an 80-degree angle). At that point it finally tears free of the water and launches itself into the air again, where of course gravity immediately begins to pull it back down.

The angle of throw is crucial. Lydéric Bocquet, a physicist at the École normale supérieure in Paris, has established that, ideally, the stone should hit the water at a 20-degree angle—that is, the leading edge of the rock should be 20 degrees above horizontal, about the same angle as a modest waterskiing ramp. That impact angle of 20 degrees minimizes the amount of time that the stone spends in contact with the water, which is crucial, because the water exerts a powerful drag on the stone, slowing it down. In general, the water is a thousand times denser than the air, so the more air time the better. A perfectly skipped stone will spend about one hundred times longer in the air than in the water.

The stone has to be thrown fast, but more important, it has to be spun fast: spinning keeps the rock stable when it hits the water. Thanks to something called the gyroscopic effect, the spinning helps prevent the rock from wobbling to one side or the other when it strikes the water's surface. Bocquet figures that a stone spinning five times a second will skip five times, but it has to spin almost twice as fast to reach fifteen skips.

 TRY THIS AT HOME! The physics behind skipping stones is one thing, but what technique is best for getting that perfect 20-degree angle and spin? Start by facing the water sideways and bending your knees. Hold the rock between your thumb and your first finger. Bring your arm back and keep the flat side of the rock parallel to the water as you throw it. At the last second, snap your wrist to flick the rock against the surface of the water.

Of course, even if your technique is perfect, every time the stone hits the water, it will give up some of its energy due to friction. Eventually, all good things must end. The final stages of a stone skip are marked by shorter and shorter skips, followed by a brief period when the rock waffles through the water without leaving the surface. These not-quite-skips are sometimes called "pitty-pats" by experts, and soon after the rock reaches that stage, it sinks.

There are annual rock-skipping competitions all over the world, and the numbers from them are startling. The most consecutive skips, as recognized by *Guinness World Records*, is an astounding eighty-eight, set by Kurt Steiner at Red Bridge, Pennsylvania, in September 2013. Steiner held several records before that particular toss and apparently spends much of his time looking for the perfect stones, which he describes as those that weigh about five ounces (King James I, take note!), measure close to one-quarter-inch thick and are extremely smooth on the bottom.

But forget the quality of the stone—the mechanics of a throw like that are extraordinary. According to Bocquet's equations, that stone had to have left Steiner's hand traveling at something like 18 meters per second, or about 40 miles per hour. Of course, this assumes a number of things about the precise angle of the throw, the actual weight of the stone and so on. But still, it would take great strength to get the rock moving that quickly, so it seems that Steiner lived up to his nickname: "Man Mountain."

As always, scientists do what scientists do. They've taken a casual activity and standardized the process by skipping perfectly round or square discs of sandstone or acrylic. That reinforces what stone skippers have figured out intuitively, but I'm with Kurt Steiner: when it comes to the proper way to skip a rock, patrol the shore, look for the perfect natural stone and let it fly.

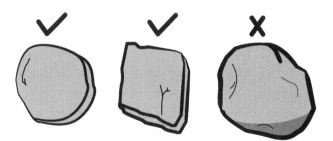

Are we living in a computer simulation?

BEFORE YOU LET THAT QUESTION STOP YOU DEAD, consider this: The Greek philosopher Plato told of people chained in a cave who saw shadows cast on the wall. The shadows belonged to puppets and the voices the people heard belonged to the puppet masters, not the shadows. But as far as the people knew, those shadows and voices were "reality." Perhaps we are no different, limited by our senses and our brains to some incomplete version of what really exists.

Cool! I'm in a book!

Today, if you've experienced high-quality virtual reality, you know that an artificial environment can be incredibly compelling. I've been in a VR version of a high-rise office where there was a balcony with no railing. When I was invited to step off into "thin air," I couldn't force myself to do it, even though I knew I was standing on a factory floor. The "reality" presented to me was simply too convincing.

And we have computer-generated alternative societies like Second Life, where you can have a perfectly normal conversation with an avatar whom you assume represents a real human somewhere, although you have no evidence for that.

Now consider the future, when computers will be immeasurably more powerful than they are today. VR and Second Life will seem like crude crayon drawings in comparison. Ray Kurzweil, inventor, futurist, and engineer, has argued that the time when computers will be more intelligent than humans will come in the next ten to fifteen years. In 2030, a computer costing a thousand US dollars will be one thousand times more powerful than the human brain.

Then imagine a time when computers are powerful enough to simulate the entire history of the earth and all the people who have ever lived on it. And they're common! Someone populates it with self-aware automatons and lets it run.

Each individual in this computer scenario would have the complete consciousness of any human today. In fact, each might actually be any human today. If you were one of those automatons, how would you know? You wouldn't. Everything in your personal past—every historical event and everything that hasn't yet happened to you but will—would simply be a computer program running its software. And you would have no idea.

You would be just another character in an amped-up version of the Sims, living your life in the equivalent of SimCity. We know—or at least are pretty sure!—that today's Sims aren't self-aware, but with much greater computer capacity, who can say they wouldn't be?

When people like inventor Elon Musk claim that "either we're going to create simulations that are indistinguishable from reality, or civilization will cease to exist," the idea is worth a closer look.

There are issues. First, will we survive to the point where we'd be capable of creating what philosopher Nick Bostrom likes to call "ancestor simulations"? It's conceivable that some sort of catastrophe might interrupt our progress toward that point, perhaps permanently. (See "Are we alone in the universe or are aliens out there?" on page 3.)

If we do achieve the kind of computing power necessary, would any citizen in that computer-enhanced world bother to do something like this? We have no way of knowing, but it would have the lure of being the closest thing to playing God that we've ever seen. (Of course, there is the idea that such a person would, in many ways, be playing the role of the many gods that have been worshipped throughout history. Gods have ultimate power over the civilizations that believe in them; computer operators would have ultimate power over their simulations.)

Science _Fiction_ or _Fact!_ It's true that real-life simulators don't bother to simulate the detail of a scene if it isn't important. They take shortcuts: maybe the people in the simulation we live in whom we think are conversing are really just saying those stock phrases uttered by movie extras to mimic conversation: "Peas and carrots, watermelon cantaloupe or rhubarb, rhubarb." Listen closely next time!

The programming of such a simulation would be a significant stumbling block. Some 86 billion neurons in the human brain, together with an even greater number of cells whose role is largely unknown, produce vivid ideas and sensations. Cells are flesh and blood; thoughts and ideas are not. How does one make the other? To give a simulation's inhabitants fully human qualities, human consciousness would have to be replicated.

If we were living in a simulation, would we know? Any computer that can simulate an entire civilization and its inhabitants would likely have safeguards against detection. But Cambridge mathematician and cosmologist John Barrow wonders if the simulation simply wouldn't be accurate enough. When Disney creates an image of light reflecting off the water, it doesn't get it absolutely right; it's just "good enough, as long as no one looks too closely." And as our knowledge increases, simulation software would have to be continually updated in the same way smartphone apps are. Then observant physicists might notice tiny mysterious changes that would alert them that something is amiss.

A usual question is: "If we're living in a simulation, how should we behave?" The straight answer is: "We have to encourage whoever's running the simulation to keep it going. If they get tired of it, it's game over!"

But how do we do that? Here are suggested tactics:

- **Live for today** (because we don't know when they might pull the plug!).

- **Be entertaining** (because that will hold their interest and prevent the plug from being pulled).

- **Try to figure out what qualities they desire in their simulated people** (in case they're the kind who want to play God).

- **Try to encourage the existence of famous and fascinating people** (we all love stories and want to be entertained ourselves).

There are people who are doing those very things. But not everyone buys it. Of simulations like this, Elon Musk said, "My brother and I agreed that we would ban such conversations if we were ever in a hot tub."

How does one pick the most private urinal in a public bathroom?

MANY MEN'S PUBLIC BATHROOMS still host a wall of urinals. There's a common belief that when men are at urinals, if they can, they will avoid standing beside another man. A space in between is so much more comfortable. So a man walks into a bathroom and is faced with a choice: Which urinal is most likely to have an open stall on both sides? How might he minimize the chances that someone will move in next to him?

A study has been done on this, believe it or not. In their paper called "The Urinal Problem," Evangelos Kranakis and Danny Krizanc take a math and computing science approach to the issue.

First, there's the obvious appeal of the two urinals at both ends of the row, because a man is guaranteed one side free of other urinators. If the bank of urinals is empty when a man arrives to the washroom, then he should probably choose a urinal at one end of the row or the other. But which end is best?

Did You Know . . . Historically, there have been cultures and moments in time when men sat to urinate and women stood. We have decided to use good judgment and refrain from including an illustration of this . . . but we did think about it.

Let's take the example of a bathroom with five urinals. If you're male, great. Proceed. If you're female, imagine you're a man. You enter the room and will probably take the closest urinal, the one at the end nearest you. How many more men will have to enter before some guy is forced to stand next to someone? If the next man chooses the middle urinal, and then the third man picks the urinal at the far end, it has taken three people to occupy all the spaces that have privacy on both sides.

Figure I

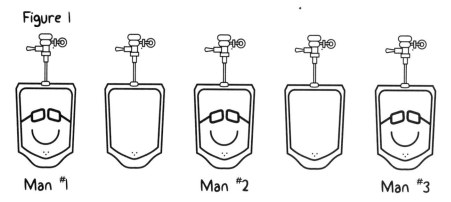

Man #1   Man #2   Man #3

But sometimes the second guy to arrive might choose urinal #4; after all, it has an empty urinal on each side. That choice messes things up, because after he picks the fourth urinal, there are no private stalls left.

Figure 2

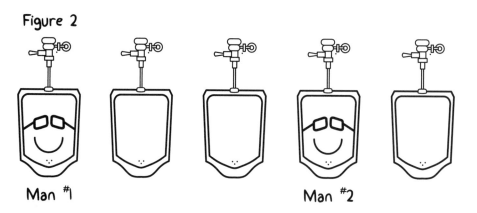

Man #1   Man #2

So, clearly the capacity for privacy depends on the second man's choice. As the first person in, though, you can ensure that there are three private stalls if, instead of choosing urinal #1, you choose #3 in the middle. That way the next two people will automatically occupy one end urinal and then the other one.

Figure 3

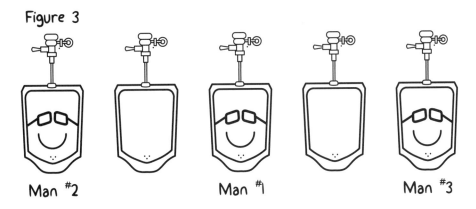

Man #2   Man #1   Man #3

Even in simple examples like this, there are other factors to consider. For instance, Kranakis and Krizanc played around with the "Lazy Filling" strategy to see how this would affect outcome. The Lazy Filling strategy refers to men's propensity to choose the urinal closest to them. If all stalls are empty, then the man will choose to be lazy and walk to the nearest stall. When there are no longer any urinals with vacancies on both sides, the next man entering the bathroom will revert to laziness and take the urinal closest to him. Because every man's bathroom goal is to find a fully private space, it turns out that if you're the first man in the urinal and want to assure your best chances at privacy even if two or three men come in after you, the urinal at the far end is the best place to head.

Kranakis and Krizanc explored many variations of the urinal problem in their paper, but taking into account every little twist and turn of human behavior makes the math of urination challenging. You might question the importance of developing algorithms to ensure that every man who visits a urinal has the greatest chance of the most private experience possible. Here is my defense: some males are so sensitive to having their personal space invaded when they need to pee that they have difficulty starting the process at all. That's the extreme, but it appears that the desire for space, as close to privacy as you get standing at a bank of urinals, is widespread among men.

In fact, there's data to back up this claim. A controversial experiment in the 1970s had as its setting a men's washroom where there were only three urinals. The psychologists running the experiment were able to limit the choices of men visiting the washroom to three options: complete urinal privacy, separated from another urinator (a man who was part of the experiment) by an empty urinal stall (with a "Don't Use" sign on it) or standing directly beside another urinator. The goal was to see if diminished privacy affected the test subject's ability to pee.

And this is where the experiment got controversial. It became clear that it was impossible to tell when a test subject had started or stopped peeing just from the sound, so the researchers arranged for an observer to sit in a toilet stall next to the urinals with books at his feet. There was a periscope in one of the books aimed at the urinals so the observer could time "start" and "stop" accurately.

The results of this bizarre experiment were as expected: the closer a test subject was standing to the man pretending to be urinating, the more time passed before the test subject started peeing—8.4 seconds, on average. But when the bathroom was empty, test subjects peed much faster: the average time was 4.9 seconds.

The ethical issues raised by spying on people as they stood at a urinal didn't go unnoticed. It's unlikely that another study involving a man with a periscope watching other guys pee will ever be replicated. But the data from the experiment fit with other well-known observations about urinal behavior. For instance, if two male friends enter a bathroom at the same time and stand at adjacent urinals, etiquette demands they keep their eyes riveted straight ahead or down, rather than make eye contact with their friend, even if they are mid-conversation. It's all about preserving the sense of personal space, even when there isn't any.

Is it me, or is there an elephant in the room?

Could we ever build a perpetual motion machine?

IF A PERPETUAL MOTION MACHINE COULD BE BUILT, it would bring the inventor untold riches and fame, and it would probably save the world, too. It's not a new idea: inventors have tried to build these machines for centuries. But no has ever been able to. No wonder: they run headlong into the laws of thermodynamics, and that's a battle that can't be won.

The first law of thermodynamics states that energy can neither be created nor destroyed. Right away, that law dooms the idea of perpetual motion. A perpetual motion machine must create energy, because by definition it produces more energy than it uses. This law doesn't mean that it's impossible to change energy from one form to another—that's something we do every day. But every time that happens, a little energy is lost in the transaction.

The second law of thermodynamics says, more or less, that energy cannot run uphill. As you sit there reading, your body is turning the nutrients from your last meal into energy to carry on the chemical reactions of life. The food molecules in your stomach are broken down, and some of the energy from that process goes toward the upkeep of your body, including generating heat to keep your body temperature around 98.6 degrees Fahrenheit (37 degrees Celsius). But you'll never recover that heat: it can't run uphill and be transformed back into the molecules that generated it. You could use your body heat to melt an ice cube you're holding in your hand, but that's still energy running downhill: liquid water can't turn itself back into ice without using more energy to do so. Energy everywhere in the universe is running down, changing from order to disorder. If the first law is sometimes expressed by the quip "You can't win," the second is "You can't even break even."

(There is a third law that deals with perfect crystals at absolute zero; even perpetual motion enthusiasts can't figure out how to defy that!)

Hey! You can't do that!

A perpetual motion machine claims to reverse this universal process and create energy where there was none before. The coolest examples are those that were designed centuries ago. One of the first was designed by Italian Marcantonio Zimara in the early 1500s. It was straightforward: Imagine a windmill built next to a giant set of bellows so powerful that, when squeezed, the force of the air emitted by the bellows would turn the windmill vanes. To complete the picture, add a third piece—Zimara called it an "instrument"—to connect the windmill to the bellows. As the windmill blades rotate, they would move the instrument, which would squeeze the bellows, then relax, then squeeze, then relax, over and over. Essentially, the instrument would let the windmill provide itself with the wind it needed to keep turning.

Zimara never built the contraption. Or, if he did, he never reported on its success, because there wouldn't have been any. The force needed to squeeze a set of giant bellows would be far greater than anything generated by the windmill, so the whole thing would just stand there motionless—not a great thing for a perpetual motion machine. The blades wouldn't even be able to turn in response to the natural wind because they would be locked onto the bellows. Zimara does get credit for being the first to design a perpetual motion machine, though.

Did You Know . . . It's not perpetual motion, but there is a machine that seems to defy physics. It's called the "EmDrive," and it's a propulsion system designed for space travel that can generate thrust without using any propellant. The EmDrive is a closed container with particles of light—photons of microwave frequencies—bouncing around inside. When the photons collide with the walls of the container, they supposedly generate a force that moves the container forward.

This sounds miraculous—and therefore unbelievable—to most scientists, but the Eagleworks research group at NASA recently published an account of their tests of the EmDrive and concluded that it did actually generate a small amount of thrust. If this turns out to be a real thing, space travel would be revolutionized, because spacecraft wouldn't have to haul massive amounts of fuel with them.

As I said, it's not perpetual motion, but it might be the cousin no one has talked about—until now.

Zimara might indeed have been the first to design a perpetual motion machine, but he was by no means the last. Robert Fludd came up with a beauty in 1618. He began with an everyday sort of mill that was common in England at the time. It consisted of a wheel—the same type that might propel a Mississippi riverboat—turned by the current of a stream, and the wheel, in turn, rotated a huge grindstone. But Fludd then introduced something called an Archimedean screw, an open cylinder with a fitted screw turning inside it. If the screw were angled upward and was constantly turning, it could carry water up with it. So, argued Fludd, all you need to do is lift the water from the river with the screw, and then let it flow along a horizontal channel and back down over the waterwheel. The turning waterwheel would be connected by gears to the screw, so the screw would keep turning, constantly lifting the water up to supply the wheel. At the same time the wheel would use its leftover energy to turn something useful, like a mill-stone, forever!

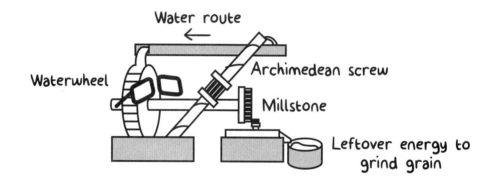

Sadly, Fludd's invention doesn't work. The work needed to lift the water is about the same as the work generated by the water turning the blades of the wheel, so there's no energy left over to turn the millstone. And that's not even counting the energy lost to the friction of the wheel turning on its axle, the screw turning in its tube or the water running along its channel. Fludd can't be blamed for not knowing the laws of thermodynamics two centuries before they were established, but inventors with designs just like this were still applying for patents a hundred years ago.

My favorite of these classic designs was based on a powerful magnet. John Wilkins, the bishop of Chester, came up with a version in the 1600s. Picture a magnet at the top of a sloping track with a metal ball at the bottom. The magnet is powerful enough to draw the ball all the way up the track. But—and this is the clever bit—there's a hole in the track just in front of the magnet, and a second track leading from the hole back to the start point. So you release the ball, it runs up the track toward the magnet, gaining speed as it does. Then the ball falls through the hole, rolling back to the start and the sequence happens again. And again. And again.

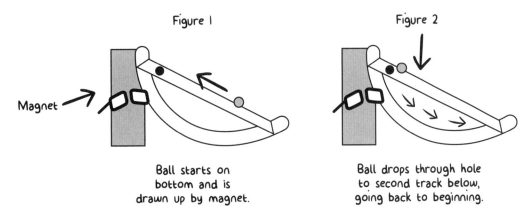

Figure 1

Ball starts on bottom and is drawn up by magnet.

Figure 2

Ball drops through hole to second track below, going back to beginning.

It's perfect. Well, perfect except for one small detail. If a magnet is powerful enough to pull the ball all the way up the track, it certainly isn't going to let the ball fall through the hole: the ball will fly right over the gap. Somebody has called this design the first executive desk toy.

Science Fiction! *In the 1870s, Monroe Paine tried to take the inventors' dream of a perpetual motion machine and turn it into a huckster scheme. He created and publicly demonstrated a new "electromagnetic machine" that he claimed could saw wood even though its power source appeared to be no more than four tiny batteries.*

A skeptic named Dr. Henry Morton suspected skulduggery, but he couldn't immediately identify it. Morton's suspicions ramped up when, during a demonstration, Paine tried to restart his machine and couldn't. Morton noted that it was 6:05 p.m., five minutes after the steam engine that ran multiple machines in the same building shut down every day. Sure enough, after Paine suddenly left town, Morton discovered a hole in the floor exactly where Paine's machine had stood—a hole just big enough to accommodate a belt drive connecting the electromagnetic machine to the building's steam engine.

So, although there have been centuries of imaginative designs for perpetual motion machines, so far the laws of thermodynamics have stood firm. One thing's for certain, though: there will be more perpetual motion machines designed in the future, using more and more exotic sources of power. All will be fascinating. None will work.

Almost-perpetual motion machine

How much do people pee in pools?

IF YOU'RE ONE OF THOSE PEOPLE who is hesitant to use the urinal in a public washroom, there's always the swimming pool or hot tub, right? Even though there's a taboo against doing that, in a 2012 survey, 19 percent admitted having done it, so you know it happens! But a percentage like that doesn't give you a clear sense of just how much urine there is in your neighbourhood pool. Well, now we know.

Urine the pool?

I am.

In a study published in early 2017, University of Alberta researchers used the artificial sweetener acesulfame K (or acesulfame potassium) as a tracer for urine. It's found in products like ice cream, jam, jelly, frozen desserts, soda pop, fruit juices, toothpaste, mouthwash, and many others. It's two hundred times sweeter than common sugar (sucrose) and is often used in conjunction with other, better-known sweeteners like sucralose or aspartame.

The beauty of acesulfame as a urine tracer is that most people consume it, our bodies don't alter it chemically as it passes through us, everyone who consumes it pees it out and it's robust enough to stick around in pool water and be detected.

Did You Know . . . That distinctive smell you associate with public swimming pools and think is chlorine? Not always. It's more often the odor of trichloramine, which forms when nitrogen in urine reacts with chlorine. Yuck!

In the study the team collected samples from two swimming pools in different Canadian cities, one with a capacity of 110,000 gallons, the other 220,000 gallons. (Olympic swimming pools are about 660,000 gallons.) The concentrations of acesulfame in the swimming pools were roughly ten times the concentrations in tap water from the two cities. The only conceivable source for this dramatic increase in concentration was urine. That meant 7.9 gallons (30 liters) of urine in one pool and 19.8 gallons (75 liters) in the other—and 19.8 gallons would fill twenty large milk jugs. That's a lot of urine!

Science Fact! There is no chemical that will turn color when exposed to urine in swimming pool water. The difficulty is identifying a chemical that would react only to pee and not other chemicals. The other roadblock is human nature: it wouldn't be difficult to imagine someone peeing in the water, then sloshing around and blaming it on someone else.

That alone has a significant "Ewww!" factor, but there's also a health implication. While urine isn't sterile, it is not generally a risk for infection, so it isn't in and of itself a problem, but the chlorine and sweat in the pool can react with it to form what are called disinfection by-products. Those can irritate the eyes and cause breathing problems, even asthma.

It should be said that respiratory problems are most often encountered by those who spend a lot of time in pools, and most swimmers like that cheerfully acknowledge that serious swimmers pee in the pool and think nothing of it. But now you know.

What's inside a black hole?

To UNDERSTAND BLACK HOLES, we first need to understand the life cycle of a sun. The beautiful thing about our sun is its consistency: just as it sets tonight, so it will rise tomorrow, and each day it will shine with the same intensity. Thanks to the crushing force of gravity keeping it in check, the sun—our closest star—has been a giant nuclear fusion reactor for billions of years and will continue to shine for about five billion more years.

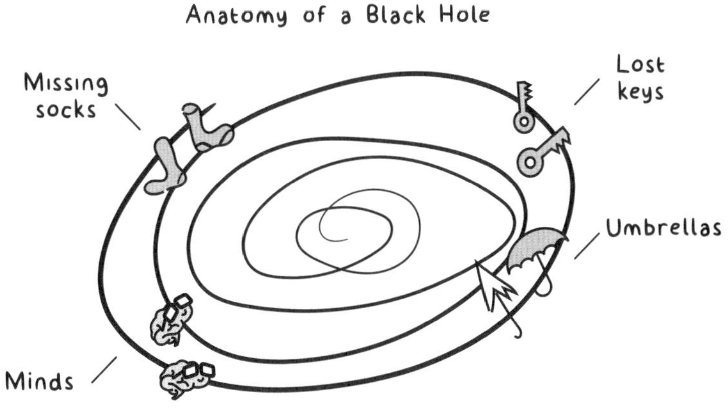

But all good things come to an end. Eventually the last of the sun's hydrogen will get used up, and its upper layers will push outward. The whole thing will heat up, swell and then shrink again. A dying sun goes through cycles like this until it sheds all of its outer layers, leaving behind what's

Anatomy of a Black Hole

Missing socks

Lost keys

Umbrellas

Minds

called a white dwarf, the dense leftover of the sun's core, or half the sun's current mass packed into an object the size of the earth.

The bigger the star, the more dramatic this dying process is. But if this story begins with a star that is even bigger—say, twenty times or greater than the mass of our sun—it will still shrink as it runs out of fuel, and the shrinkage never stops. The gravitational force of that much mass will simply continue to collapse until it reaches an incomprehensible state of infinite density packed into an infinitely small space. This is a black hole.

There are black holes that are star sized, some that are supermassive and, theoretically, mini black holes. Each one has the same mass it did before it collapsed (minus any material that was shed into space), so there must be something in them. Black holes, especially the huge ones, vacuum up stars and gas that they encounter. As that material swirls down the drain of the black hole, it heats up and emits huge quantities of X-rays and radio waves. From the pattern of those objects, astronomers can determine the size of the black hole that they're orbiting. The late physicist John Archibald Wheeler described this process as similar to standing in a poorly lit ballroom where the male dancers are dressed completely in black but their female partners are in white. You would be able to make out only the female dancers, but you could judge by the movements of the females where the male dancers were.

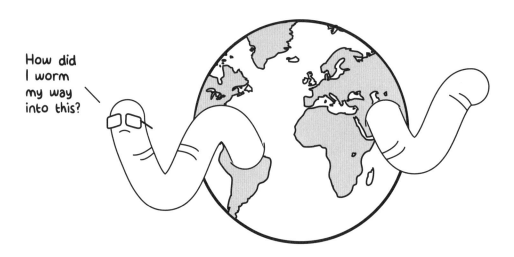

How did I worm my way into this?

TRY THIS! A black hole's closest relative is a wormhole. Wormholes are both exits and entrances in space—places that you can (hypothetically) move through to reach distant parts of space much more quickly than you could had you traveled the conventional way. Understanding wormholes requires some mind stretching. First, imagine space not as a three-dimensional area but as a two-dimensional sheet of rubber held between your hands. A massive object, like the sun, has the same effect on space as a billiard ball dropped onto the rubber sheet: it warps it. Then, with the billiard ball in the middle of the sheet, take a smaller ball—say, a marble—to represent Earth, and roll it across the sheet. It won't travel in a straight line. It will gravitate to the middle and travel around and around the billiard ball (representing the sun) in an orbit because of the slope of the sheet.

Now imagine using a pen to make two ink marks at opposite ends of this rubber sheet. Now fold the sheet to position the two points on top of each other. Once the points overlap, you can either travel from one to the other by sticking to the surface of the sheet and tracing a path all the way around, or, instead, you can punch your way right through the rubber: that's the wormhole. Congratulations, you just traveled in hyperspace. Just don't make the mistake of thinking you can travel through a wormhole by entering a black hole: that would be the end of you!

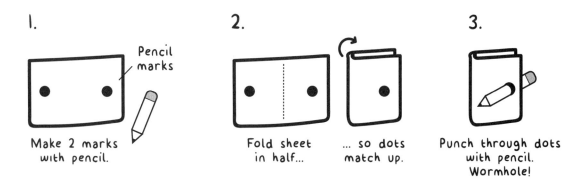

1. Make 2 marks with pencil. Pencil marks

2. Fold sheet in half... ... so dots match up.

3. Punch through dots with pencil. Wormhole!

We can't know anything about a black hole beyond what's called the event horizon. This is the border surrounding the black hole. Once you cross it you can't ever come back, because the black hole's gravitational force is so strong. If you did cross it, your experience would be very different from those watching you. Outsiders tracking you as your spaceship descended toward the event horizon would see you get closer and closer, and then you would appear to stop. The light of your image wouldn't be sucked into the black hole, but it wouldn't be able to escape the force of the hole's gravity, either; it would remain there, suspended in space against the darkness of the black hole.

As the person flying into the black hole, though, you would experience events much more quickly. What makes a black hole special is not that its mass is unbelievably huge but that the mass is packed into such a tiny space—a single point—and so its gravitational effect is much more intense. On the earth, because your feet are slightly closer to the center of the earth than your head, you are technically experiencing an infinitesimally stronger gravitational force on your feet. But the distance between your head and feet is so small relative to the distance to the center of the earth that you don't feel the effects.

A black hole is a much more concentrated source of gravity, not to mention more massive than the earth. As you get closer to the black hole's center, you would begin to be stretched. That can't last for long, of course, because you'll start to break apart, first into a top half and bottom half, then into four, then eight, and so on. At the same time, space-time itself is being channeled into the black hole, meaning that you're being squeezed from the sides, as Neil deGrasse Tyson likes to put it, "like toothpaste." It's actually more like the last gasp of the toothpaste when it comes out in splatters—that's what it would be like for you to enter the black hole. This simultaneous stretching and squeezing has a technical term: spaghettification.

After you enter the black hole in skinny little pieces, it's unclear what exactly would happen, because we have no way of looking inside. The "black" in the name means that no light escapes from the intensity of the hole's gravity, and where there's no light, we can't see what's going on. What we know for sure is that the laws of physics (as we know them) no longer hold. Maybe space simply ends there and the story's over. But as far as we know, matter shouldn't be able to end its existence, so that doesn't seem right. Although there's an infinite density squeezed into a tiny space, there's a chance that your remnants—even the fragments of your atoms—could survive. Another idea, which few people believe but is hard to disprove, is that your existence would somehow end in a wild shower of particles. It seems that, no matter what the ultimate fate is of an object entering a black hole, it would violate the laws of quantum mechanics and/or general relativity, and physicists are very reluctant to part with either one.

Spaghettification (and meatballs)

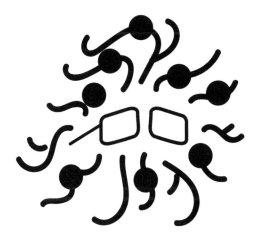

Will machines ever have feelings?

"STOP, DAVE . . . WILL YOU STOP, DAVE . . . Stop Dave. I'm afraid . . . I'm afraid, Dave . . . Dave . . . my mind is going . . ."

That's HAL, the computer in Arthur C. Clarke's (and Stanley Kubrick's) *2001: A Space Odyssey*. HAL, a machine with feelings, agonizes as astronaut Dave begins to take him apart. He's reacting with emotion—with terror, in fact. Without emotion, HAL would have been nothing more than a mindless number-crunching machine.

I'm feeling binary today, just like yesterday.

He's fictional, though. And while we're nowhere near being able to build a machine like HAL, could we one day? And if we did, would that mean the machine with feelings would be more intelligent than all those without?

Here's some food for thought. In 1848 a man named Phineas Gage was construction foreman on a crew building the Rutland & Burlington Railroad in Vermont—routine work, putting blasting powder into a hole. But on September 13, everything went sideways: the powder exploded, and a rod drove right through Gage's skull and landed 65 feet (20 meters) away. The rod took out his left eye and much of the left frontal lobe of his brain.

Incredibly, Gage survived the accident and lived for many years afterward. But people who had known him before testified that "good guy" Gage was gone. No longer friendly, even-tempered and reliable, he became a bad decision maker and a sociopathic drifter. The terrible wound to his brain had transformed him.

More recently, American neurologist Antonio Damasio described a patient of his, Elliot, whom he referred to as a "modern-day Gage." Elliot had a brain tumor surgically removed, and parts of both his left and right frontal lobes were removed as part of the operation. Afterward, Elliot suddenly started making one bad decision after another. By the time Damasio saw him, he had lost virtually everything, including family and wealth. Oddly, Elliot passed myriad psychological tests, but he confessed that in many situations in which he used to have feelings, he no longer felt anything. Damasio became convinced that Elliot, like Gage before him, had lost an essential component of sound decision making: emotions.

Sometimes I can't see the forest for the trees.

Even lab studies have shown that people in happy moods are better able to focus on all aspects of a drawing and later reproduce it than sad people are. The happy people focus on the whole forest and can recreate the full picture; the sad ones focus on specific trees and therefore literally and figuratively fail to see the bigger picture. That suggests that a robot with "moods" might make better decisions.

But can we prove that emotionally intelligent machines might be more capable than those without feelings? And if so, how? The challenge is that human feelings are the product of millions of years of evolution, and the circuits of brain cells that produce them are very poorly understood—they're so complex. So right now, anyway, imitating them is out of the question.

Taking baby steps first, how about creating a computer that could at least recognize human emotions? Researchers in China have shown that a computer can identify positive, neutral and negative emotions from the brain waves of humans watching film clips. Then there's the French/Japanese store robot, Pepper. It can recognize four emotions: happiness, joy, sadness and anger. Pepper then responds in a way that might encourage shoppers to buy more. Still, these robots and computers are responding automatically; they have no idea what people are actually feeling. Even if we build computers that are able to interpret any human emotion, they still have to be able to respond appropriately.

Universal Cat Translator

Meow.

Love is a many-splendored thing.

Mee-ow.

Your human smell calls me to scratch your eyes out.

Mrreow.

Nietzsche was right. And I feel nothing.

Meeew.

Do androids dream of electric sheep?

The emotions you experience—your feelings—are inaccessible to anyone unless you express them. You might look angry to me, but that doesn't mean you actually are angry, and I have no way of probing your consciousness, that inner world of thoughts, dreams and feelings.

Consciousness is a scientific mystery: not only do we not know how a brain generates consciousness, we don't know which living things have it. Crows and ravens are capable of insightful actions, but are they conscious? If dogs and cats are self-aware, that's evidence that a human-sized brain wouldn't be a prerequisite for a robot. Some scientists have argued that once a brain reaches a certain level of sophistication, consciousness might spontaneously emerge. Would that be true of a complex assemblage of microprocessors as well?

Did You Know . . . IBM is trying to replicate the interconnectedness of the human brain. TrueNorth is an IBM computer chip that simulates neurons—16 million of them, which in turn make 4 billion connections. That's impressive, although 16 million can't compete with the 86 billion in the human brain.

If robots were able to develop emotions, would they even be like ours? Maybe they'd have "machine feelings": no "heart," just a battery. A machine with feelings might be a more effective machine, but it might also be much less predictable. And if all those sensitive but smart robot types decide they don't need us around anymore . . . then what?

Did You Know . . . Scientists in Germany showed people a computer-animated scene of a man and a woman discussing their feelings about the hot weather, their lack of spare time and the woman's annoyance at having been stood up by a friend. Onlookers were told either that the voices were human or computer-generated. Those who were told the conversations were computer-generated and unscripted thought the scene was eerie. Those who were told these were human interactions did not. What does this mean? Apparently we feel that there's something disturbing about a computer that is thinking entirely on its own. But we might just have to get used to that.

What's lurking in your bathroom?

THE BACTERIA THAT LIVE IN OUR BODIES, also known as our microbiome, have the ability to positively influence what goes on in our bodies, even in our brains. But we don't really keep those bacteria to ourselves all that well. Almost anything that we touch, or breathe on, or even come close to, will be covered with them: computer keyboards, phone mouthpieces, your socks—anything.

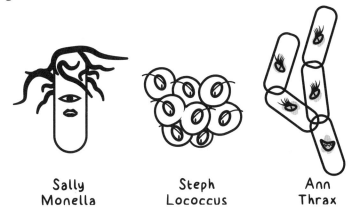

Sally
Monella

Steph
Lococcus

Ann
Thrax

Creating a map of how bacteria spread is complicated but extremely important, because it's crucial information for managing food- and water-borne illnesses. Important rest stops on that map include the places humans pee and poo (washrooms) or bathe (the shower) and, in particular, the doorknob, the toilet handle, the toilet seat and even the showerhead.

There's a ton of research on people's behaviors in public washrooms. (See "How does one pick the most private urinal in a public bathroom?" on page 131.) For instance, women wash their hands much more often if they think someone is watching them, and men avoid certain stalls due to the practically imperceptible whiff of a human male hormone.

The latest research gathered in the public washroom has allowed us to create a complete catalog of bacteria that live there. Looking at the results, bacteria aren't on the endangered species list! The survey found more kinds of bacteria in public washrooms than there are birds in North America. And scientists are familiar with them all, because they're bacteria we already have on our skin or in our gut. In the washroom, just as in our bodies, the places those bacteria colonize are different.

The research used groundbreaking methods. Traditionally, a cotton swab is used to transfer bacteria to a petri dish, which is covered with nutrient gel so the bacteria can multiply. Scientists can then study the colonies of live bacteria on the plate to see what they're dealing with. But hundreds of species are incapable of growing on a dish like that. Instead, for the North American survey, the samples were examined to see what DNA traces had been left from different species. The researchers never saw the bacteria, just their entrails.

Doorknobs, soap dispensers, taps, and other things touched by hands were revealed to be rich in skin-dwelling microbes. But toilet handles and seats were much different. There, gut bacteria predominated: somehow fecal matter was making it all the way from people's guts to those surfaces. That can be attributed to poor hygiene or eruptions of fecal-rich material when a toilet flushes. Since serious disease-causing bacteria can spread through poo, the fact that you find fecal bacteria in lots of places where you shouldn't means their pathogenic cousins could survive there, too.

Did You Know . . . You wouldn't notice it, but a toilet flush flings lots of fecal bacteria into the air. Those bacteria can survive for days wherever they settle. So if you're at home, put the lid down before you flush.

It was no surprise to discover that the underside of the toilet bowl rim in public washrooms carried heavy bacterial loads, but what was surprising was that the floor hosted the richest variety of germs. Researchers suspected that part of that was because soil from different places was being tracked in, but, oddly, the same array of bacteria species was also found on toilet handles. That provided evidence that the apparently common germophobe habit of flushing the toilet handle with shoes instead of hands was alive and well.

The study concluded that if toilets are used all day every day, bacteria from pee and poo will inevitably spread throughout the washroom, so it makes good sense to wash your hands thoroughly—for your own sake.

And here I thought I was getting clean!

When it comes to the showerhead, the bacterial content gets particularly interesting. Given that the showerhead is above us, we can't drip germs onto it. Even drops that bounced straight back into the showerhead would carry only bacteria from our scalp—and how likely is that? There's still the possibility that flushing the toilet could make its inevitable contribution toward the showerhead's germ population, but that's not the whole story.

Did You Know . . . Oligarchs are extremely powerful, rich, apparently untouchable individuals. Populations of bacteria that succeed in dominating places are also called oligarchs. They have the right mix of defenses, tolerances and advantages that allow them to run roughshod over competitors.

Interestingly, belly button bacteria separate themselves into oligarchies. The most surprising of those mini-despots is a specimen belonging to the domain Archaea. Those are microbes that were once thought to be the only organisms that could live in extreme environments, such as hot springs, but have now been discovered almost everywhere. They had never been found before on the human body until they were detected in belly buttons.

The microbial population of the showerhead is taken seriously because some research has revealed that showerheads are the most likely part of the bathroom to have a group of bugs known as the *Mycobacterium avium* complex. While these germs make up a fraction of 1 percent of the bacterial populations in city water supplies, their ability to form stable films on plumbing equipment means they dominate in showerheads. Worse, showerheads dispense bacteria in tiny airborne drops that move about and can enter your lungs, and the mycobacteria can cause lung disease in people whose immune systems are compromised. For most people, though, those bacteria don't pose a significant risk.

I have to admit, it's ironic to write about the proliferation of bacteria in environments that we created to eliminate them. But let's be serious and think about contamination before the flush toilet. How about the good old days, when people just poured waste into the castle moat? Or, if you couldn't afford a castle, down the walls and onto the street? And, of course, for a big chunk of human history, people didn't bathe much. What we face now with toilets spewing bacteria into the air and odd bugs gathering secretly in showerheads sounds bad but isn't.

News flash: put the lid down
before you flush me!

What is the Turing test?

ALAN TURING WAS A BRILLIANT ENGLISH MATHEMATICIAN, code breaker and computer scientist who deciphered the German Enigma code, but he is probably best known for creating the Turing test, a way of deciding if a machine is intelligent. As such, he could be considered the father of AI, artificial intelligence.

The test, as Turing originally designed it, was straightforward, even though his goal was profound: "I propose to consider the question, 'Can machines think?'" It took only a few more sentences before he realized that such a straightforward goal would run into problems, given the commonplace and ambiguous definitions of both "machine" and "think." To solve this complicated question, he picked a relatively simple approach: he chose to create a variation of the party game called the "imitation game."

I have a mind like a steel trap.

Imagine this situation: A man and a woman sit in one room, a judge in another. The judge asks questions of both (in print, preferably typewritten) and tries to figure out, based on each person's answer, which is the woman. That's the party game. Turing tweaked that idea by substituting a computer for one of the people. The judge had to decide from the answers which was the human and which was the computer. Turing noted that the answers didn't have to be correct, only that they resembled an answer that a human might give.

Turing refined the game by saying the computer would pass the test if the human judge could do no better than accurately identify either human or computer more than 70 percent of the time, after a five-minute conversation. Turing also allowed space for future development by asking whether, if the computers of his age failed the test, "imaginable" computers could win the game. He thought the test would be won by the year 2000.

In the paper he wrote describing the test, Turing raised a number of possible objections to it, then took out each one like someone shooting ducks at an arcade. When people argued that God gives souls to humans, but not machines, Turing said, "I am not very impressed with theological arguments." And when some people said that intelligent machines were too scary a thought, he said, "I do not think this argument is sufficiently substantial." He also dismissed arguments that machines have limits, that machines cannot write poems and therefore can't think and that machines can't originate anything.

Did You Know . . . The 2014 movie *The Imitation Game* provided a glimpse into the mind and life of Alan Turing. Turing was a gay man at a time when gay relationships were considered a criminal act. He was outed in 1952, tried and convicted. He avoided jail by agreeing to chemical castration, but died two years later from cyanide poisoning at the age of forty-two. It was assumed to be suicide, although there have been suggestions that it was accidental. Regardless, as attitudes modernized, Queen Elizabeth granted Turing a posthumous pardon in 2013.

The original version of the Turing test has yet to be officially passed, but several people claim to have beaten it. In the mid-1960s, Joseph Weizenbaum created a program called ELIZA, which was designed to mimic the conversational efforts of a style of psychotherapy called "person-centered" or "Rogerian" therapy, after the inventor Carl Rogers. The job of a Rogerian therapist is to encourage patients to reveal themselves through questions. So comments such as "How does that make you feel?" and "Does that surprise you?" are legitimate but don't really require the machine to do a lot of thinking. Weizenbaum claimed victory, although his claim has been disputed ever since.

Since then, varieties of the test have sprung up. One of my favorites is the Chinese room. Like everything in this area of research, it's contentious, but is a neat example of the kind of thinking surrounding ideas not just of machine intelligence but consciousness in general.

In this scenario, a person who knows no Chinese whatsoever is locked in a room full of information about Chinese symbols. There is also a book with directions about how to arrange these symbols in the proper order. Chinese speakers outside the room pass questions in Chinese into the room, and the person inside—without knowing that these are questions—consults the book of instructions, puts together a set of symbols and passes them outside—again, not even realizing that the set is the answer to a question. As far as the Chinese speakers outside are concerned, there is someone (or something) inside the box that is passing the Turing test. But since the person inside knows absolutely nothing about the Chinese language, they can't be said to be passing any sort of test of intelligence.

Two more of my favorites are the reverse Turing test, in which a computer has to tell if it's conversing with a person or another computer, and the total Turing test, where the computer being tested also must see and handle objects—artificial intelligence plus machine vision plus robotics.

I have trouble thinking outside the box.

Whether you know it or not, you've likely played a modified version of the reverse Turing test yourself. Any time you go to a website that requires you to type a set of distorted letters and numbers in a box, you are trying to demonstrate to the computer monitoring you that you are human. It's called CAPTCHA: Completely Automated Public Turing Test to Tell Computers and Humans Apart. Computers aren't nearly as good as humans at deciphering the CAPTCHA letters and numbers. Humans, on the other hand, are so good at making sense of unclear images that we can see the face of the Virgin Mary in a grilled cheese sandwich!

Did You Know . . . The Loebner Prize, created by Dr. Hugh Loebner, offers $100,000 and a gold medal of Loebner himself to the first entrant who passes the Turing test. It was started in 1991, but even with the progress in computing since then, it has yet to be won. Smaller prizes are given out to what you might call "best in show" each year, but none have reached the bar Turing set.

While writing this, I chatted with the chatbot deemed the best in the most recent Loebner competition, and it seems we still have a long way to go: when I mentioned dinner (no, I wasn't asking for a date!), it went off on a tangent about Charles Dickens. Maybe it was thinking of "Please, sir, I want some more" from his novel *Oliver Twist*.

History Mystery

What is the Antikythera mechanism?

We are so technologically sophisticated that we hardly spend any time marveling at how far we've come. But every once in a while we're brought up short by a discovery that shows us that we have always been technologically minded. The ultimate example of this is something few people have heard of, let alone are familiar with: the Antikythera mechanism.

Imagine a clock made of metal and set in a beautifully carved wooden case, about 13 x 7 x 3.5 inches (34 x 18 x 9 centimeters)—about the size of a small, flat breadbox stood on end. This clock has two large concentric dials on the front, a handle on the side for winding, and more clock-like faces on the back. Inside, there are thirty different gears crammed in, all interlocked. Some of them connect directly, others in a way that one gearwheel turns slower than the other, and some in an off-center way like a Mad Hatter teacup ride on a midway: think of following the movements of our moon as it orbits the earth while at the same time tracking the earth as it orbits the sun. (Gears like this weren't seen in Greece for another 1,500 years!) Everywhere across this device are elaborate inscriptions. It is incredibly complex, beautifully designed, and it's close to 2,000 years old.

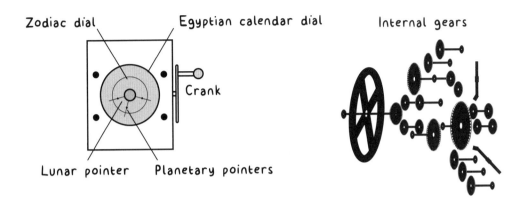

The Antikythera mechanism got its name because, for the longest time, it wasn't clear what it was supposed to do. It was discovered by chance in 1901 when sponge divers in the eastern Mediterranean were driven ashore by a storm near the small Greek island of Antikythera. The next day, when the seas were calm again, they resumed diving and discovered the wreck of a Greek ship about 150 feet down that dated back to roughly 100 BCE. Together with Greek archaeologists, they recovered a wealth of pottery and jewelry. In many ways, it was the first serious underwater archaeology operation.

Did You Know . . . Three thousand four hundred characters have been discovered on the Antikythera mechanism so far, but there were perhaps twenty thousand covering it originally. Some of the original written surfaces are missing; sometimes the inscriptions are readable only by seeing their imprint on the stuff that was caked on it.

Among the wealth of artifacts were three small pieces of encrusted, corroded brass so devastated by centuries of exposure to seawater that there were almost no traces of gears and dials. While it was clear that the pieces belonged together somehow, it was a mystery what they might have formed.

The analysis of the mechanism began decades later with X-rays, but the remains were so crushed and distorted that the scans didn't provide much information. The most recent analyses used microfocus X-ray computed tomography and revealed sensational aspects, including faint traces of gear teeth no longer in existence. From those images, it's been possible to reconstruct what the mechanism might have looked like.

The two dials on the front represented the 365-day calendar and the 360-day calendar of the Zodiac. The handle on the side enabled the user to set the device for a precise day—in the past or the future—either to see planetary alignments in the past or the future. A gear-driven calendar isn't all that impressive, but the dials on the back were a different story. They were spirals, not circles, and they tracked the movements of the five planets that were known at the time to an accuracy of 1 degree every five hundred years. Not only that, this mind-boggling piece of machinery was even able to track the sun and moon in precise detail as well as predict lunar and solar eclipses, events that might not have even been visible from the eastern Mediterranean.

Did You Know . . . I can tell you confidently that in the year 2034 there will be a full moon on Christmas Day. The last time that happened was 2015. No, I'm not Nostradamus or some sort of wizard. Thanks to today's science, the movements of the sun and moon are known in such detail that I can make that prediction without a shred of doubt. But even without all of our modern technology, you could have made the same prediction using the Antikythera mechanism, because it was capable of keeping track of the 235 months that it takes for a full moon to reappear on the same day of the calendar.

That's not as simple as it sounds. Even though it takes a little less than twenty-eight days for the moon to complete one orbit around the earth, the time from full moon to full moon is a bit longer. A full moon requires complete illumination by the sun, and in the twenty-eight days of the moon's orbit, the earth moves enough that it takes an extra day or two for the moon to be in the right position to be fully illuminated.

The timing is tricky to measure, but the Antikythera device was so finely tuned that it could even track what's called the Saros cycle, the 6,585 days and 8 hours necessary to account for an eclipse (either lunar or solar). And, yes, the mechanism even tracked those pesky eight hours.

And if all that wasn't impressive enough, the Antikythera mechanism tracked social events, too! One of the displays on the back looked ahead four years to the next Olympiad while also acknowledging five lesser-known games that happened every two years.

 **Did You Know . . .** In 2013 the watchmaker Hublot created a "simplified" wrist version of the Antikythera mechanism, a one-off. While it is not for sale, it would certainly cost millions if it were.

More than a century's worth of research has made it clear that the Antikythera mechanism was an incredible piece of design and craftsmanship that wasn't equaled for another thousand years: there is no other Greek artifact of the time—or even hundreds of years later—that uses gears with such precision.

It might have been intended for a temple or some incredibly rich individual. It could have been a sports calendar, a horoscope, or something else entirely. One thing that most certainly remains, though, is a mystery.

Science _Fact!_ In ancient Greece, there were predictions of the color of eclipses and even the winds that would accompany them, as those were factors thought to influence people's lives. Some people believe, then, that the Antikythera device was meant to mix astrology with astronomy.

There's no way to be certain either of the color of an eclipse or of the winds it might bring, so there's not much scientific evidence to back up those astrological claims. But although Greek artisans are credited with having built the Antikythera mechanism, the astronomy that it measured originated from Babylonian ideas that arose hundreds of years before the mechanism was created. That means that Greek knowledge about planetary movements might have been inspired by the gear wheels and movements of the device. So it was a case of technology inspiring theories, not the other way around.

Acknowledgments

THIS BOOK IS A FOLLOW-UP to *The Science of Why*, a book that in many ways was novel for me: a book of answers to questions, complete with cartoon drawings to add humor, a book that combined a casual and approachable look with science. I was hoping that it would have broad appeal (fourteen-year-olds to eighty-four-year-olds), and it seems to have accomplished that. And now *The Science of Why²*.

It has been great working with my old friend Kevin Hanson at Simon & Schuster Canada (especially over the occasional sushi lunch), and also Nita Pronovost, my editor, who somehow manages to shape a manuscript without appearing to. Either she's a very skilled editor or she casts a spell over the writer—or both. But it works for me. There are many others at S&S who help take a book from concept to bookshelves, too many to mention, in fact, but I would acknowledge Catherine Whiteside and her publicity team for their earlier work generating awareness and even excitement about *The Science of Why*. I know I can count on the same for this book.

I thank those scientists who took the time to expand on the subjects I was addressing.

They include: Rachel Kulik, Greg Kawchuk, Christophe Demoulin, John Hutchinson and Kory Czuy.

I also had fantastic research help and commentary from Joanne O'Meara, a physics professor and science communicator at the University of Guelph. Physics threads its way throughout this book, and she was there to help me negotiate the trickiness of it all.

Thanks too to my agent Jackie Kaiser and all the people at Westwood Creative Artists in Toronto.

One of the crucial tests of whether an idea is worth writing about or not is the reaction it generates among friends. If a casual mention is greeted with a raised eyebrow or a dismissive laugh, or both, that's usually a good thing. I have a good cast of such people, including the guys in the Beakerhead Band, the people at Beakerhead themselves, Niki Wilson in Jasper and the group who wander off to Montana every summer. Thanks to you all.

And the CEO of Beakerhead, Mary Anne Moser, is the one who maintains her interest in projects like this, despite having way too much on her mind already. Not just interest; creative interest. Couldn't do without it.

Photo: Richard Siemens

Jay Ingram has written thirteen books, including the bestselling first book of this series, *The Science of Why*. He was the host of Discovery Channel Canada's *Daily Planet* from the first episode until June 2011. Before joining Discovery, Ingram hosted CBC Radio's national science show, *Quirks & Quarks*. He has received the Sandford Fleming Medal from the Royal Canadian Institute, the Royal Society of Canada's McNeil Medal for the Public Awareness of Science and the Michael Smith Award for Science Promotion from the Natural Sciences and Engineering Research Council of Canada. He is a distinguished alumnus of the University of Alberta, has received six honorary doctorates and is a Member of the Order of Canada. Visit Jay at JayIngram.ca.

🐦 @jayingram

Looking for more head scratchers and mind benders?

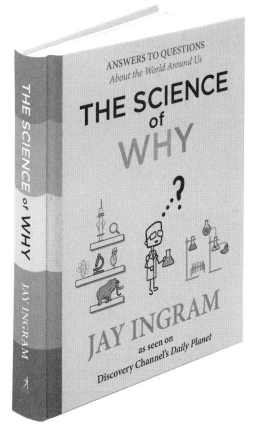